# ARDUINO MANUAL IN ENGLISH TOMO III

# SENSORS

More fun to do is to use LEDs flashing all sensors to detect what is happening "out there" and react accordingly. Unfortunately, each sensor has its own methods of connection: some need resistance "pull-up" and some not, some need their own sources of power and some not, some work to much stress and not others, etc. In this chapter the most common sensors are presented with examples of circuits that are used and Arduino code that make them work.

He also indicated for each specific type of sensor which specific products can be found in different distributors. However, if desired, can be purchased easily once a set of different sensors thanks to "sensor pack 900" of Adafruit (code Product No. 176) or the "Sensor Kit" Sparkfun (product code 11016 ). The first includes an infrared LED and a specific infrared remote sensor, a light sensor, a temperature sensor, a tilt sensor, shock sensor (usable as buzzer), magnetic field sensor (with a magnet ), a force sensor and an accelerometer. The second includes a specific infrared remote sensor, a light sensor, a bending sensor, a sensor for shock and vibration, magnetic field sensor (along with a sensitive switch-what he called a "reed switch "-), a force sensor, a humidity sensor, a distance sensor, a motion sensor, an accelerometer, a gyroscope, a compass (magnetometer) and an atmospheric pressure sensor (barometer). It also includes a thin membrane potentiometer with linear path (product number 8680).

Another interesting sensors kit is provided by Cutedigi with product code H21 which contains a temperature sensor, humidity, sound, Hall effect, tilt, obstacles, fire, metal, an accelerometer, one compass, LDR a "reed switch" ... plus an infrared transmitter and receiver, a button, a buzzer, an LED RGB a optointerruptor, and more.

# VISIBLE LIGHT SENSORS

## Photoresists

Light sensors, as its name implies, are sensors that detect the presence of light in the environment. Sometimes they are called "cells CdS" (by the material of which are usually made, cadmium sulfide) or also "photoresists" and LDRs (for "Light Dependent Resistor") as basically consist of a resistance changes depending on the amount of light that is incident on its surface value. Specifically, they are reducing their strength as they receive more light intensity.

Usually they are small, cheap and easy to use; They appear so much in toys and household devices in general. But they are imprecise: each photoresist reacts differently to another, although they were manufactured in the same tier. That is why they should not be used to determine exact levels of light intensity, but rather to determine variations in it, which can come from ambient light itself ("dawn or dusk") or the presence of any obstacle blocking reception of any incident light. It could also have a system of photoresists and so compare which one gets more light at a given time (to build such a follower of roads painted white on the floor or light bulbs, among many other applications robot).

Another thing you need to know is your typical response time is on the order of a tenth of a second. This means that the change in its resistance value has that delay regarding changes in light. Therefore, in circumstances where the light signal varies quickly use it is not indicated (although it is also true that this slowness in some cases is advantageous because rapid changes in lighting and filter).

When purchasing a photoresist must also take into account a number of other factors besides the size and price, above all we must also look at the minimum and maximum resistance that might offer. These data can be obtained from the datasheet provided by the manufacturer. In fact, in the datasheet we can not only see these two extremes data but also all intermediate values of resistance, thanks to a set of graphs showing us how varies continuously (and usually logarithmic) the resistance value of the photoresistor based on the amount of light received, measured in lux units. With this information we can know, knowing the light falling on the LDR, which provides the resistance value (and conversely, knowing the resistance offered, we can deduce the amount of light received by the sensor).

In the foregoing paragraph a "set of graphic" rather one is mentioned because the behavior of the photoresist is dependent on ambient temperature as this, the resistance variation with respect the incident light will be either. It is also mentioned that this variation is "generally logarithmic"; there may be cases where it is not, but what is certain is that by the very nature of photoresists itself, the relationship between amount of light received and resulting resistance will never be linear. This means for example that if the illuminance is 10 lux and a resistance of 100 $\Omega$ is measured, when this is

20 lux resistance will have to be 50 $\Omega$ .

Another point to check on the datasheet to take account of the sensitivity. The photoresists detected likewise the different types of light; specifically, they are usually more sensitive to changes in light green than in other colors. There also exist a minimum wave length (400 nm, typically) and maximum (600 nm, typically) beyond which do not detect anything. This information can be found in the form of graph showing the response of the photoresist depending on the wavelength received.

Brief note on the electromagnetic spectrum:

Call "light" waves are electromagnetic type. Therefore, as waves that are one of its features it is that we can study their frequency

as measured in Hz- or alternatively its wavelength as measured in nanometers

(10-9 meters) -. Both variables are related by the expression $\lambda = c / v$

(Where $\lambda$ is the wavelength in vacuum, the frequency $v$ $c$ is the speed of light in a vacuum is a constant equal to 299,792,458 m / s).

We can classify different types of light by wavelength and have gamma rays and X-rays (with shorter wavelength), passing continuously (! light is an analog quantity) by ultraviolet light, light visible and infrared rays, reaching electromagnetic waves of longer wavelength as are radio waves. The set of all these waves is called the electromagnetic spectrum

The human eye is unable to perceive the entire electromagnetic spectrum, only a small part identified as "visible light". This means that in our daily lives, we are surrounded by electromagnetic radiation really do not see. Although no exact limits for the visible region of the spectrum (depends on each person) are usually taken as accepted values waves having a wavelength between 400 nm and 700 nm.

Within the visible spectrum, depending on the particular wavelength having a specific wavelength, this will be of one color or another. So, we can say about that visible light of wavelength between 400 and 450 nanometers is violet, between 450 and 495 blue, between 495 and 570 green, between 570 and 590 yellow, 590 to 620 orange and between 620 and

700 red.

Furthermore, the voltage accepted by these devices can be virtually any (to 100 V). Because photoresists are just resistors are not polarized, so their terminals can be connected to our circuits in both directions.

The easiest way to check that a light sensor function is to connect its terminals to a multimeter resistance measurement mode and make an impact more or less light. If we see that responds (monitor with scaling), and we can begin to design our projects with him.

The first thing you have to know is how a light sensor is connected to our circuit. What we will do is connect one of the terminals of the power sensor and the other through a "pull-down" resistor (about whose proper resistive value will discuss a few paragraphs below), grounded. In addition, from a point between the photoresist and the "pull-down" resistor connect a "third wire" to an analog input of our Arduino so that it can read the analog voltage to be measured. This voltage received may range between 0 V and the

feed the photoresistor voltage: the following figures show the photoresistor fed with the 5 V pin "5V" Arduino, but could also feed perfectly with the 3.3 V pin "3V3", for example.

If we use a "pull-down" resistance, the higher voltage through analog input receive the Arduino mean more light strikes the sensor. If we had used a resistance "pull-up" would be the opposite: the higher the voltage received mean that there is more darkness. We, as mentioned, use a resistance "pull-down" so that, ultimately, the assembly would be similar to this:

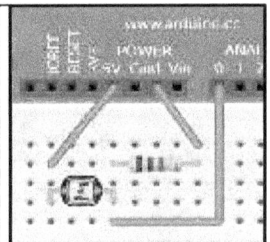 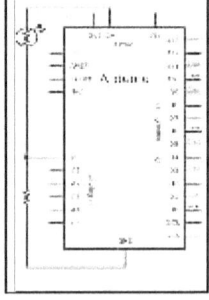

The trick to understanding in depth this show is to See That as the resistance of the photoresistor is decreasing (Because It Affects more light), the overall resistance of resistors in series photoresistor + pull-down Also decreases. By Ohm's Law, This Will make (to keep fixed supply voltage across the circuit-from 5 V) the current intensity Increases THROUGHOUT the circuit. But as the "pull-down" resistor is fixed, for the same Ohm's Law, the current passing-through if it has Increased, it has done so the voltage across terminals STIs. That is indeed what we measure With the "third wire" existing voltage Between the terminals of the "pull-down" resistance. If the resistance of the photoresistor out to be much light- -at negligible, would be the Measured voltage 5 V; if the resistance of the photoresistor was so great to open the circuit-at disrupting the passage of electrons be much darkness, the Measured voltage would be 0 V. Between These two extreme cases Have intermediate Measures.

As Explained in the PRECEDING paragraph can be Summarized in the following formula, derived from Ohm's law and the fact That the intensity That Runs Through Both resistors is the same: Vavg = (Rpull / (Rpull + Rfoto)) · Vfuente, Where Vfuente is the voltage supplied by the power supply, the voltage Vavg is received by the analog input pin (ie, the one Between the terminals of the resistance "pull-down", Which can be worth Between 0 V and Vfuente) it is the value pull -up resistor "pull-down" (fixed) and Rfoto is the resistance value of the photoresistor. From here you can see what We Have Already Said That Increases When the light intensity (as the resistance of the photoresistor decreases), the voltage Also Increases Measured and vice versa.

However, in reality is not the value we work Vavg With our Arduino, Because It always use an analog to digital for mapping of all values received analog converter (Which can range from 0 V and

Assuming That the 5 V provided by the source voltage is 5 V) to digital values (ranging Between 0 and 1023). These digital values are what Arduino Those Who Actually Understand and work in our sketches. The Good News Is That the conversion of analog to digital can be Expressed by the naked rule of proportionality THUS: Vconvertido = Vavg · 1023/5. From here, if we substitute esta expression in the formulated above, and cleared away Rfoto, arrived at the following expression: Rfoto = (Rpull · 1023 / Vconvertido) - Rpull, Which Allows us to know finally what the current value of the resistance of photoresistor from the digitized voltage Obtained by the Arduino.

The next step naturally, eleven Known the value of the resistance of the photoresistor, would find out how much lighting is Appropriate. This

You can see in the graphs of the datasheet, as with previously discussed. However, This step is not done Because Usually the photoresists are normally used to compare lighting (between sites or Between different time) rather than specific values for lighting.

Appropriate To find the value of resistance "pull-down" We Have to put in our circuit, we should know (from the datasheet) different specific numerical values can purchase our photoresistor along various lighting (Rfoto) and use, Along With the value of the pull-up in the Formula Already Obtained seen in the previous page Vavg = (Rpull / (Rpull + Rfoto)) · Vfuente to see what we would get Vavg hypothetical. By doing This, That we see in most cases, the behavior of Rfoto received Regarding lighting Makes Rpull high values (for example, 10k) Quickly saturate the readings in bright environments. That is, the sensor Makes measure Reaches the top of the 5 V with a Relatively low light and not reliable to distinguish THEREFORE Between a well lit another very well-lit environment. However, lower values of pull-up (for example, 1 kW), it will allow detecting Changes in the brightest light but will not be Able to distinguish Differences in dark levels. THEREFORE, Depending on the environment Where we place our project, we choose one of 10 k pull-up (for dark environments) or 1 k (for bright surroundings), or use some type of adjustable potentiometer.

Example 7.1: Ensure and how to photoresistor behaves when connected to an Arduino board. So we can assemble the following circuit; Set the LED must be connected to a PWM pin.

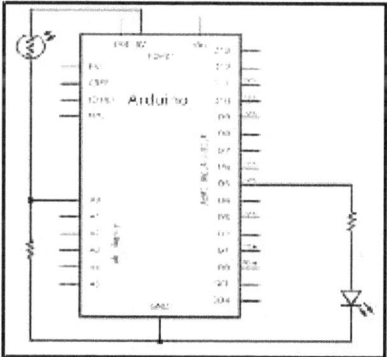

What we want is to use the voltage value read on pin analog input (No. 0 in this case) to illuminate consequently the LED: the less light detected by the photoresistor, brighter light LED (hence it is necessary that the LED also receives an analog signal through a PWM pin, # 5 in this case). Also it is shown by the serial channel the values obtained by the photoresistor.

```
int valorcds;  //Valor obtenido
int brilloLED;  //Valor enviado al LED
void           setup(void)           {
        Serial.begin(9600);
}
void loop(void) {
        valorcds = analogRead(0);
/*Además de imprimir " valorcds"  tal cual, también se podría haber comprobado si este es
menor o mayor que una cantidad dada, y haber imprimido un mensaje tal como " Oscuro" ,
" Normal" , " Brillante" , etc*/
        Serial.println(valorcds);
```

/*El valor obtenido " valorcds" será mayor cuanto más brillante sea el entorno. En cambio, el LED ha de iluminar más cuanto más oscuro sea el entorno. Es decir, " brilloLED" ha de ser mayor cuanto menor sea el valor de " valorcds" . Por eso, hemos de invertir " valorcds" para que su valor pase de una escala de 0 a 1023 a otra de 1023 a 0.*/

```
        valorcds=1023-valorcds;
```

/*Y ahora, tal como ya hemos visto en ejemplos anteriores, hemos de mapear " valorcds" para que caiga dentro del rango admitido para la salida PWM. Es decir, pasar un valor que está entre 0 y 1023 ha otro que está entre 0 y 255.*/

```
        brilloLED = map(valorcds, 0, 1023, 0, 255);
        analogWrite(5, brilloLED);
        delay(100); //Para que se pueda ver el nuevo brillo
}
```

With what is already known, we could perform a simple presence detector. If we kept illuminated the photoresistor steadily, to interpose an obstacle between the light source and the photoresist, this would detect a sudden drop in light intensity. Using the same circuit from the previous example, we could slightly modify the sketch to send a digital output signal to the LED, so this was worth HIGH (lighting the LED) if the photoresistor will detect a value below a certain threshold value chosen by us (and therefore detect someone comes between light and sensor) or worth LOW (turning off the LED) if the value "valorcds" is greater than the threshold (and therefore a "normal" incidence is detected light sensor). It is left as an exercise.

On the other hand, we have said before that the readings obtained by a photoresistor as we use for our projects (for example, Product No. 9088 Sparkfun or Adafruit # 161) are not very accurate. It is highly recommended therefore calibrate these components before you start to work with them (for example, within the "setup ()" function. Calibration is setting the minimum and maximum range of possible values to read if you already know that these never reach 0 or 1023, respectively. In this way can better interpret the readings because the intermediate values are more consistent.

Example 7.2: The following sketch aims to calibrate a photoresist, although the procedure is almost unchanged generalizable to any analog sensor. The circuit required is the same as in the previous examples: A photoresistor connected to an analog input pin (assume 0) accompanied by a "pull-down" resistor 10 k, and an LED connected to an output pin PWM (assume 5) accompanied by their voltage divider 220 corresponding Ω. The sketch is basically what makes reading during their first five seconds of running a series of values of the analog sensor to establish what their reading with minimum value and what will be the one with the maximum value. Clearly, we submit during those five seconds, the sensor both exigent circumstances to see how he reacts (in case of a photoresist, illuminating light with maximum foreseen in the project and with the minimum). After the calibration, the rest of the code is very similar to that seen previously: the key is in the map (function), first because she performs mapping ranges invested (as already explained to illuminate the LED when it detects low light), but mostly because the mapping is established in the range of calibrated values, not the usual 0 and 1023.

```
int valorcds = 0;
int  sensorMin  =  1023;  //Irá  disminuyendo  int
sensorMax = 0;   //Irá aumentando void setup() {
        //Calibramos durante los primeros 5 segundos de programa while (millis() <
        5000) {
                valorcds = analogRead(0);
        /*Si se lee un valor mayor que el actual máximo, lo guardo como
           el nuevo valor máximo */
                if (valorcds > sensorMax) {
                        sensorMax = valorcds;
                }
        /*Si se lee un valor menor que el actual mínimo, lo guardo como
           el nuevo valor mínimo */
                if (valorcds < sensorMin) {

                        sensorMin = valorcds;
                }
        }
}
void loop() {
                valorcds = analogRead(0);
```

/*Aplico la calibración a la lectura recién leída a la transformación que ha de sufrir valorcds para ser usada en analogWrite() */

    valorcds = map(valorcds, sensorMax, sensorMin, 0, 255);
/*En el caso de que la lectura recién leída caiga fuera del rango establecido durante la calibración…*/

        valorcds = constrain(valorcds, 0, 255);
        //Ilumino el LED usando el nuevo valor calibrado analogWrite(5,
        valorcds);
}


7.3 Example: If we have the circuit shown in the figure below (ie, an LDR resistance "pull-down" of 10 k and a buzzer with a voltage divider 100 $\Omega$ , for example), can make it as light detected by the LDR, the frequency of the sound emitted by the buzzer go changing.

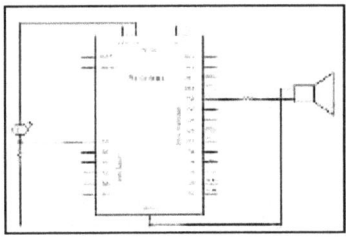

The code is this: simply makes reading the LDR, the maps to a range of values and sends audible buzzer. It should first be calibrated to adjust the LDR mapping conveniently as map values () of the sketch are only approximate.

```
void setup() {}
void loop() {
        int lectura;
        int sonido;
        lectura = analogRead(0);
/*En este caso, la calibración del LDR da valores entre 400 y 1000, pero esto puede variar.
El rango de salida son, en Hz, las frecuencias mínimas y máximas del sonido que
queremos emitir*/
        sonido = map(lectura, 400, 1000, 120, 1500);
        tone(9, sonido, 10);
        delay(1); //Por estabilidad entre lecturas
}
```

It would not be too difficult to replace in the above circuit the LDR by a potentiometer. In fact, this same variant we saw in the section on the sound of Chapter 6.

Example 7.4: LDRs Another practical example is the design of a device capable of moving towards the darkest point of the environment, thanks to the values obtained by two strategically placed light sensors. The idea is to compare the reading of both sensors and guiding a servomotor in one direction or the other depending whether the sensor reading is higher or lower than the other. The assembly would be something like this:

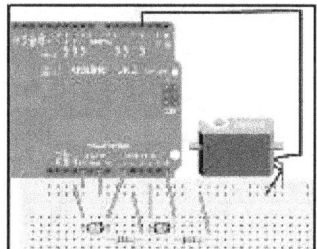

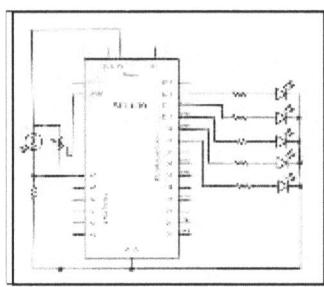

And the code, this:

```
#include <Servo.h>
int   valorLDRderecha  =   0;   int
valorLDRizquierda = 0; int angulo =
0;
Servo miservo;
void setup() {
    miservo.attach(3);
}
void loop() {
    valorLDRizquierda = analogRead(0);
    valorLDRderecha = analogRead(5);
    if (valorLDRderecha < valorLDRizquierda) {
            angulo = angulo –  10;
            if (angulo < 0) {angulo = 0;}
    } else {
            angulo = angulo + 10;
            if (angulo > 179) {angulo = 179;}
    }
    miservo.write(angulo);
}
```

A suitable value for the voltage dividers of the LEDs may be 220 Ω . A suitable value for the "pull-down" resistance associated with the LDR can be of 1 kW. A suitable value for the maximum resistance of the potentiometer can be 10k. The LDR as you can see, is connected to the analog input pin number 0.

The idea is to illuminate our environment as the LEDs go dark. But instead of using a single LED regulated analogously as we saw in one of the above examples, now we use several LEDs controlled by digital signals so that more LEDs as they are detecting more darkness will be illuminated.

The novelty of this circuit is the use of a reference voltage controlled by a potentiometer. The reason for including this element is as follows: we know that the reading obtained from the photoresist can range from 0 (corresponding to 0 V, when the room is completely dark) to 1023 (corresponding to 5 V, when the environment is fully lit). But these two extreme cases can be difficult to get, so maybe just go to work in a range of 500 to 600 intermediate values so much contempt resolution. To fix this and to further allow our circuit suits extreme environments with different lighting without changing the threshold values written in our code each time, it was decided to use an adjustable voltage reference. This voltage marks the point at which, if the ambient light decreases, begin to decrease the values obtained in the analog input pin. If this reference is low, the LEDs start to glow in dim light but will be more sensitive to slight variations in illumination; if this reference is high, the LEDs will begin to glow with bright ambient light but this will vary greatly to modify the number of illuminated LEDs. In other words, the potentiometer is used to set the threshold of minimum light, from which begin to run our circuit artificial light.

This is because varying the reference signal, we are saying that the range of values 1024 are located between 0 V and a given maximum voltage that will be one or the other depending on the circumstances. In our sketch we have divided the 0-1024 range into five sections which progressively activate each of the five LEDs. If our environment offers a very low variation light to play with, we can adjust the reference voltage to a low value (for example, 1 V) and thus we will continue to distribute proportionally the activation of outputs, so we get a higher sensitivity. The "price to pay" is not activate the circuit until the analog input does not receive those 1 V, since while receiving a higher voltage, the analog-digital converter is saturated.

14

With this trick, then, we will not have to make any changes to the code if we change to an environment with changes in lighting more or less extreme: just by varying the position of the potentiometer, the range 0-1024 automatically adapt to the new external lighting circumstances. The code is as follows:

```
int valorLDR = 0;
byte i;
void setup() {
  //Utilizaremos 5 LEDs, conectados a los pines 8,9,10,11 y 12 for (i=8;i<=12;i++) {
  pinMode(i,OUTPUT); }
  analogReference(EXTERNAL);
}
void loop() {
  valorLDR = analogRead(0);
  /*Si el entorno está muy iluminado (según la referencia marcada por el
    potenciómetro), no se enciende ningún LED */
  if(valorLDR >= 1023){
          for (i=8;i<=12;i++){ digitalWrite(i,LOW); }
  //Si está algo menos, se enciende un LED
  } else if(valorLDR >= 823){
          digitalWrite(8, HIGH);
          for (i=9;i<=12;i++){ digitalWrite(i,LOW); }
  //Si está algo menos, se encienden dos LEDs
  } else if(valorLDR >= 623){
          for (i=8;i<=9;i++) { digitalWrite(i,HIGH); }
          for (i=10;i<=12;i++) { digitalWrite(i,LOW); }
  //Si está algo menos, se encienden tres LEDs
  } else if(valorLDR >= 423){
          for (i=8;i<=10;i++){ digitalWrite(i,HIGH); }
          for (i=11;i<=12;i++) { digitalWrite(i,LOW); }
```

```
//Si está algo menos, se encienden cuatro LEDs
} else if(valorLDR >= 223){
        for (i=8;i<=11;i++){ digitalWrite(i,HIGH); }
        digitalWrite(12,LOW);
//Si el entorno está muy oscuro, se encienden los cinco LEDs
} else {
        for (i=8;i<=12;i++){ digitalWrite(i,HIGH); }
}
}
```

# The digital sensor TSL2561

Besides the photoresistors (which are analog sensors), there are light sensors are digital, such that the chip TSL2561 Adafruit breakout distributed over a comfortable insert. Digital sensors are more accurate than the photoresistors (since they allow accurate readings, measured in lux units) and sensitivity can be set depending on the light intensity with which work at that time (intensity which permitted range is also much wider than the photoresists). Furthermore, TSL2561 specifically, besides the visible spectrum, also detects infrared light; gases and separately measuring visible light, infrared light, or both.

This chip is supplied with a voltage of between 2.7 V and 3.6 V and works at most 0.5 mA, making it ideal for low-power systems. Their communication with the outside is the I2C protocol, so the breakout plate in which, in addition to the power contacts and ground markets contacts "SDA" appear (to connect to No. 4 analog pin of Arduino ) and "SCL" (connect to analog Arduino pin # 5).

The accuracy of this sensor has a "price", and is difficult to use: in addition to the internal complexity itself that provides the I2C protocol, to figure the exact amount of light read by the chip are of little use many mathematical calculations intuitive. Fortunately, Adafruit Arduino library itself offers a greatly facilitates the collection and interpretation of data, downloadable from https://github.com/adafruit/TSL2561-Arduino-Library. For lack of space we can not delve into the use of this library, but if you want to start learning how to use it after installing it as any bookstore Arduino, I recommend watching the commented on the sketches of example that come along with the library code.

The analog sensor TEMT6000

There are also chips light sensors with analog behavior. One example is the TEMT6000 distributed as per Sparkfun breakout plate (product # 8688). This sensor has the advantage of being much more accurate than a photoresistor (better react to lighting changes in a wider range) without adding more complexity to our circuits. It is adapted to the human eye sensitivity, which does not react to infrared or ultraviolet light.

18

Connections distributed by Sparkfun breakout plate are very simple: the "VCC" connector can be connected directly to the pin "5V" Arduino, the "GND" connector ground "and connector" SIG "to any analog input Arduino. The more we read voltage sensor will be brighter environment.

Example 7.6: The programming is also very simple. Here's an example:

```
int pinsensor = 0; //Entrada analógica donde está conectado el sensor void setup() {
    Serial.begin(9600);
}
void loop() {
    int lectura;
    lectura = analogRead(pinsensor);
    Serial.println(lectura); //0=muy oscuro; 1023=muy iluminado delay(100);
}
```

Another breakout plate that includes the same chip TEMT6000 Freetronics is distributed under the name "Light sensor module". It also has three connectors (VCC, GND and OUT) to communicate with our Arduino and program exactly the same.

Another breakout insert very similar (with three connectors also programmable in the same way) is called "Light sensor AMBI" Modern Device, but this includes another chip, the GA1A1S201WP.

19

## SENSORS INFRARED LIGHT

### Photodiode and phototransistor

A photodiode is a device that, when

excited by light, the circuit produces one (measurable) proportional current circulation. Thus, they can be used as light sensors, although if it is true that there are particularly sensitive to visible light photodiodes, the vast majority are mainly infrared light. They can be purchased from any distributor of components

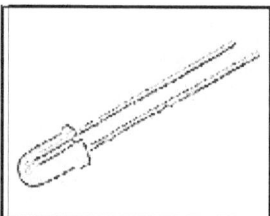

basic, such as Mouser and Jameco, to name a couple of them. Examples of specific devices that we can come in handy are (the code is the manufacturer) the TEFD4300F, the BPV22F, the BPV10NF or SFH235FA.

Keep in mind that, despite having a behavior similar to the LDRs appearance, an important difference from these (in addition to the sensitivity to other wavelengths) is the response time to changes in dark to light, and vice versa, which is much smaller photodiodes.

As with standard diodes, photodiodes have an anode and a cathode, but attention, desire to function as a photodiode always be connected to the circuit in reverse polarity. Course, as with ordinary diodes, usually the anode is longer than the cathode (in case of equal length, the cathode must be marked in some way).

Its internal operation is as follows: when the photodiode is biased directly, the light striking it has no appreciable effect and therefore the device behaves as a common diode. When it is reverse-biased and does not get any light radiation, also it behaves as a normal diode as the electrons flowing through the circuit does not have enough energy to cross it, so that the circuit remains open. But the moment in which the photodiode receives light radiation within a range of appropriate wave length, the electrons get enough energy to "jump" barrier diode in reverse and move on.

Example 7.7: To test their behavior, we can use a circuit like the next page. This circuit is identical to what we saw with the LDRs, replacing these by a photodiode (which is identified by a new symbol that we had not seen before). The value of the voltage divider depends on the amount of light (infrared) present in the environment: greater resistance improve sensitivity when there is only one light source and less resistance the better when there are many (the sun itself are sources or lamps infrared); a value of 100k can do well to start. Notice that also is the cathode of the photodiode (the shorter terminal, remember) which is connected to the supply.

The operation of this circuit is as follows: while the photodiode not detect infrared light, the analog input of the Arduino board (in this case the number 0) 0 V voltage is measured because the circuit will act as an open circuit. As you increase the light intensity on the photodiode, increase the amount of electron transfers it (ie, the intensity of
current). This implies that, when the resistance "pull-down" fixed, Ohm's law the voltage measured at the analog input pin will also increase, reaching a moment in which to receive much light the photodiode not only cause resistance passage of electrons and therefore Arduino read a maximum voltage of 5 V.

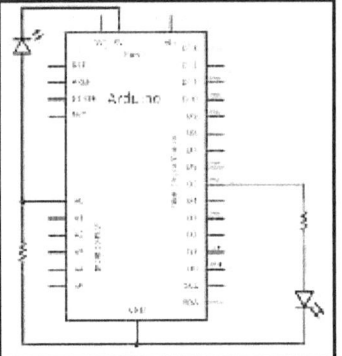

We've added a LED connected to PWM output pin # 5 as we did when we saw the LDRs to have a (pun intended) visible way to detect the incidence of infrared light on the photodiode. As can be seen in the code used (shown below), we made dependent intensity LED brightness of the amount of infrared light detected by the photodiode: the more infrared radiation is received, brighter is the LED.

```
int valorfotodio;     //Valor obtenido del fotodiodo int brilloLED;
//Valor enviado al LED
void setup(void) {
        Serial.begin(9600);
}
void loop(void) {
        valorfotodio      =      analogRead(0);
        Serial.println(valorfotodio);
        /*El brillo del LED es proporcional a la cantidad de luz
         infrarroja recibida */
        brilloLED = map(valorfotodio, 0, 1023, 0, 255);
        analogWrite(5, brilloLED);
        delay(100); //Para que se pueda ver el nuevo brillo
}
```

We can play to bring the photodiode in the previous example to different sources of infrared light (almost any relatively hot object functions as such), but it is best to add to the previous source circuit infrared something more manageable. The most common use for it is emitting infrared LED. Simply connect its anode to pin 5 V Arduino and its cathode to ground (through a voltage divider 220 Ω is fine) we would have a steady and stable source of infrared radiation. If we wanted a source of controllable issue, we would like to connect more than any other LED: its anode to an output pin of the Arduino (digital or PWM as we want) and its cathode to ground (through a divider Stress also, of course). Sparkfun distributes an infrared LED emitter 850 nm with product number 9469 and another 950 nm with product number 9349. Adafruit distributes a 940 nm LED with product number 387.

Another type of light sensors besides photodiodes are called phototransistors, ie light sensitive (also typically infrared) transistors. Its operation is as follows: light to impinge on its base, it leads to a current transistor to a conducting state is generated. Therefore, a phototransistor is equal to a common transistor with the only difference that the base current Ib is dependent on the light received. In fact, there phototransistors that can work both ways: either as phototransistors or as common transistor with a current Ib given base concrete.

The phototransistor is much more sensitive than the photodiode (by the effect transistor gain itself), as the currents that can be obtained with a photodiode are really limited. In fact, a phototransistor can be understood as a combination of photodiode and amplifier, so, actually, if we build a home phototransistor, it would suffice to add a common transistor photodiode connecting the cathode of the diode and the collector of the transistor the anode base. In this configuration, the current delivered by the photodiode (which circulate towards the base of transistor) is amplified β times.

In many circuits we can find a phototransistor within walking distance of an infrared LED emitting a wave length compatible. This pair of components is useful for detecting the interposition between them of an obstacle (due to interruption of the light beam) and thus act as optical switches. They can be used in many applications, such as detectors of passing a credit card (at an ATM) or the introduction of paper (to a printer) or tachometers, among many others. A tachometer is a device that counts the revolutions per minute that performs an obstacle subject to

a wheel or blade rotates (usually due to the operation of a motor); that is used to measure the speed of an object.

We can acquire and factory under a common encapsulated couple more phototransistor with the generic name of "photointerrupter" LED components. In Sparkfun for example distributed with each code No. 9299, which has two terminals for the anode and cathode of the LED, and two terminals for the collector and emitter of an NPN phototransistor. Usually, we want to connect the terminals of the LED to a continuously fed closed circuit (anode source cathode to ground), the collector terminal of the photointerrupter for a power supply and the emitter terminal of the photointerrupter to a digital input of our Arduino, to thereby detect the occurrence of current when light is received. On the other hand, both the input Arduino and the issuer should be grounded through the same resistance "pull-down" for more stable readings (a typical value of 10 k may work, but depending on the circuit might need higher values).

23

We can also find more infrared LED phototransistor pair in a component called "optocoupler" or "opto-isolator". Its schematic representation is usually as follows:

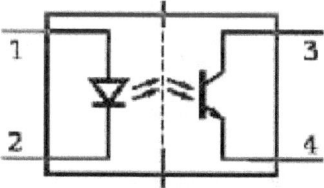

Roughly an optocoupler acts as a closed circuit when light comes from the LED to the transistor base and open when the LED is off. Its main function is to control and simultaneously isolate two parts of a circuit that normally work at different voltages (just as you would a common transistor, but a little more securely). Physically often chips that offer at least four legs (like photointerrupters): two corresponding to the terminals of the LED and two for the collector and emitter of phototransistor (although they may be more appropriate to base leg if allowed to control the intensity this also flowing as standard). Examples are the optocoupler 4N35 or CNY75, manufactured by several companies available at Mouser, Jameco and the like.

The LED-phototransistor pair is also useful for detecting objects at small distances from it. This will be discussed in the corresponding section distance sensors.

## Remote control

An immediate practical utility of a transmitter-receiver pair infrared (such as an LED and a photodiode / phototransistor) located at a distance is sending "messages" between them. That is, since the infrared light is not visible (and therefore not "bothered") can emit pulses of specific duration and / or frequency that can be received and processed several meters away without "note" . The device that receives must then be programmed to perform different actions depending on the type of read pulse. Sparkfun for example sells a "pack" with both components, product number 241.

In fact, any device that works with a "remote" works similarly on the front because I have to have an infrared sensor (also called sensors "IR" English "infra-red") that receives signals Infrared issued by the command. And what's inside This is basically an LED that emits pulses of infrared light following a pattern that points to the device the order to realize: there is a code to turn the TV blinks, another to change channels, etc.

In the previous paragraph we talk about "IR sensor" and not photodiode / phototransistor because the former are more sophisticated. Specifically, the IR sensors do not detect any infrared light, but only that which (thanks to incorporating an internal bandpass filter and demodulator circuit) is modulated by a carrier wave of a frequency of 38

KHz + 3 KHz. This basically means only signals whose information is carried by a wave of 38 KHz will be read. This is to prevent

IR sensors will "go crazy" when receiving the infrared light coming from all sides there (sun, electricity ...) thus only respond to very specific and standardized emissions.

Another difference with photodiodes / transistors is that the IR sensors offer a binary response: if they detect an IR signal from 38 KHz value that can read them in most cases is LOW (0 V), and if not detected nothing, reading offers a HIGH (5 V) value. This behavior is what is often called "active low, or" low-active ".

Examples of IR sensors can be TSOP32838 Vishay (Product No. 157 or No. 10266 Adafruit Sparkfun) or GP1UX311QS. The most outstanding features are its sensitivity range is between wavelengths

800 nm to 1100 nm with a maximum response at 940 nm and need about 5 V and 3 mA to operate. SparkFun also markets (with product code

8554) a very simple breakout plate with another IR sensor, the chip TSOP85.

The TSOP32838 chip offers three tabs: having to face the hemispherical back, the pin on the far left is the digital output provided by the sensor, the center pin must be grounded and pin on the far right must be connected to food (between 2.5 V and 5.5 V).

Example 7.8: To test its performance, could design a circuit like this. The voltage divider for the LED may be between 200 and 1000 ohms could have also added a voltage divider between 100 and 900 Ω in series with the supply pin sensor and also a bypass capacitor 1 uF connected between the ground and to stabilize the behavior of the sensor, but not essential.

The idea is to turn the LED briefly when the IR sensor detects an infrared signal. But attention, not just any infrared signal, but only that modulated at 38 KHz. Therefore, to test this circuit, we can not use an infrared LED either: we use a remote control we have at hand (from a TV, a DVD player, a

computer, etc.). Once loaded in the Arduino sketch then presented, if we aim for that command to the IR sensor and press some of its buttons, we should see that the LED lights. In this way, we will be using the IR sensor like a switch, which lights up the LED while detecting that signal and turns it off when you no longer detected.

The sketch takes for the circuit to operate as we wish above is as follows (the output of the sensor is connected to digital input No. 2 of the Arduino and LED board is connected to the digital output No. 12):

```
int        irPin=2;        int
ledPin=12;           void
setup() {
        pinMode(irPin,INPUT);
        pinMode(ledPin,OUTPUT);
  }
  void loop() {
  / * Since the signal emitted by the sensor is normally HIGH when a control button is pressed, it
  switches to LOW. What makes the pulseln () function is pausing the sketch until a signal
  LOW, the duration really does not interest us but obviously always be greater than zero is
  detected. Therefore, if the condition Is true means it you pressed a button on a remote control
  * /
  if (pulseln (Irpin, LOW)> 0) {
```

/ * You must expect a certain time (depending on the specific remote control) following the detection of the first LOW signal because each button press produces multiple oscillations between HIGH and LOW values. Although physically it has nothing to do, we can understand this waiting like a way to avoid a "bounce" (phenomenon studied when trying buttons). Once after this timeout, the sensor signal should be returned to its resting state (HIGH value). * /

```
                delay(100);
/*Mantenemos encendido el LED durante unos cuantos milisegundos. Durante este tiempo el
sketch no podrá detectar otras pulsaciones provenientes del control remoto. También
podríamos haber enviado un mensaje al " Serial monitor"  notificando la pulsación. */
                digitalWrite(ledPin,HIGH);
                delay(200);
                digitalWrite(ledPin,LOW);
        }
}
```

However, beyond here we can not do much with our Arduino without knowing the specific patterns of pulses emitted by the IR remote control used. That is, to prevent a Sony remote control to change channels on a Philips TV (for example), each brand uses a different encoding for its signals, although all are modulated at 38 KHz. That is, each brand emits different durations, sequences and combinations of HIGH and LOW signals, to avoid possible interference. Therefore, if we want to control an Arduino board using a remote control of any device we have at hand a particular brand, we should first know the protocol used for this so that our plate to process properly. And conversely: if we want to remotely control an appliance of a given brand by an Arduino board (which would in this case remote control) should know the communication protocol recognized by the appliance. Unfortunately, there are almost as many patterns as equipment IR pulses, but we have several solutions for it.

The first would buy a couple more control IR remote sensor that would be compatible factory without having us to worry about its internal communication protocol. In this sense, we can acquire the "IR Kit" DFRobot (Product No. DFR0107), which includes a remote control and IR sensor supports (with product number DFR0094 if purchased separately). This sensor comes in a breakout plate with three pins (5V, GND and digital signal) that can easily connect to our Arduino or a breadboard. Interestingly, this kit comes with a library of downloadable web product that allows our Arduino properly interpret a data received by the digital signal pin IR sensor.

Another similar alternative is to acquire, in Adafruit, the pairing of the remote control product number 389 (we do know that NEC uses the pattern of signals) and TSOP38238 IR sensor used in the previous example circuit . If used with the remote control indicated, this sensor can be controlled by a library of Adafruit itself, "Adafruit NEC Remote Control Library", downloadable https://github.com/adafruit/Adafruit-NEC-remote-control -library

SparkFun in turn distributes a complete kit with product number 10783, which includes a remote control (with product number 10280 if purchased separately), a pair of IR sensors also TSOP38238 model plus a pair of IR LEDs also. In order to process signals from the remote distributed in this kit (and only that particular command) correctly, there

an official bookstore, but product page offers us an Arduino code examples where use of a simple function itself getIRkey use call () is made, which returns the numeric code corresponding to the button pressed at that time (or 0 if not pressed any). On the other hand, with the same remote control, you can also use the breakout plate with product number

8554 mentioned a few paragraphs above; in this case, yes you can use a

library, available in https://github.com/konstantint/ArduinoSparkfunIRReceiver.

A different approach to the previous solutions for the remote control of our project is to not use any remote control, but construírselo yourself. It is easier than it may seem a priori, as to emit infrared signal modulated at 38 KHz only thing you need is a simple infrared LEDs. Yes, it is necessary to control it so that it can turn on and off at 38 KHz many times as necessary. That is, if we emit an infrared LED that lasts a HIGH signal (for example) a second and remains lit LED just during that time and that's it, the IR sensor does not detect anything because the signal is not modulated. To modulate the signal HIGH LED will flash on and off 38,000 times during that second. Thus, the IR sensor will recognize that frequency pattern and interpreted that a pulse has arrived. If we issue a LOW signal, to keep the LED off and it would.

Thinking this application markets a breakout even Sparkfun plate (with code 10732) which includes an infrared LED and a transistor so that the emitted light has a greater intensity and more easily grasp at greater distances. It is only necessary to send the modulated signal to the base of the transistor (corresponding to connector plate labeled "CTL"), in addition to feeding said plate and grounded.

Example 7.9: The following is an example code for issuing a dummy pattern modulated pulses at 38 KHz, repeated infinitely shown every second. This pattern is formed by pulses of varying duration, separated by time intervals of different duration also. This is intended to simulate the pattern of any commercial product, in which pulses of different lengths are usually appear at different intervals. We assume that the LED is directly connected to the digital output pin No. 8 Arduino (and ground through a voltage divider, as always).

```
void setup(){
      pinMode(8, OUTPUT);
  }
  void loop(){
      pulseIR(2080);//Emito un pulso modulado durante 2080 µs delay(27);    //No emito
      nada durante 27 milisegundos pulseIR(440); //Emito un segundo pulso modulado
      durante 440 µs delayMicroseconds(1500);//No emito nada durante 1500 µs
      pulseIR(460);
      delayMicroseconds(3440);
      pulseIR(480);
      delay(1000);
  }
  /*Esta función envía un pulso modulado a 38KHz de una
  duración determinada por su parámetro.*/ void pulseIR(long
  microsecs) {
      cli();
      /*Voy pasando los periodos que " caben"
      en el intervalo de tiempo especificado */
      while (microsecs > 0) {
```

/*Una frecuencia de 38KHz equivale a un periodo de 26 microsegundos: durante 13 microseconds (semiperiodo) la señal ha de ser HIGH y durante 13 microseconds será LOW */

```
        digitalWrite(8,HIGH);
/*Es conocido que la ejecución de la función digitalWrite tarda sobre
3 microsegundos, por lo que tras ella espero 10 microsegundos más */
        delayMicroseconds(10);
        digitalWrite(8,  LOW);                      //Se  tarda  también  3  µs
        delayMicroseconds(10);      //Me espero 10 µs más
        microsecs = microsecs - 26;  //Ya se ha cumplido un periodo
    }
    sei();
}
```

Cli () and sei (): In the above code Arduino two functions of language that we had not seen before appear. The cli () function is used to disable certain tasks Arduino is always constantly occurring in the background (such as listening possible data received by the serial port, keeping track of time, etc.). When treating high speed signals, it is advisable to keep the Arduino as inactive as possible to make an accurate and clean up. The second serves to return activate these tasks.

Example 7.10: It would not do to have a transmitter pulse modulated if we have a sensor that recognizes them. To do this, we use the IR sensor TSOP38238 (with its output connected to pin # 2 of the Arduino board) and the following code, which displays the Serial monitor the pulse duration modulated received (HIGH values), and the existing time including (actually, the duration of the LOW pulse). With this code, in fact, we can learn the patterns of any commercial remote control and therefore we can build us a clone:

```
//Pin de la placa Arduino donde está conectada la salida del sensor const byte irPin=2;
//Duración máxima que reconoceremos para un pulso (¡65 ms es mucho!)
const int MAXPULSO = 65000;
/*Resolución temporal que utilizaremos. Cuanto menor sea su valor, más precisa será la
lectura de la señal, pero si es demasiado pequeño puede haber problemas de sincronización.
El valor de 20 suele ser el adecuado en la mayoría de los casos*/
const byte RESOLUCION = 20;
```

/*Guardaremos 100 parejas formadas por un pulso (valor HIGH) y su valor LOW siguiente. Para ello hacemos uso de un array bidimensional: el valor HIGH de la primera pareja se guardará en el elemento pulsos[0][0], el valor LOW de la primera pareja se guardará en el elemento pulsos[0][1], el valor HIGH de la segunda pareja se guardará en el elemento pulsos[1][0], el valor LOW de la segunda pareja se guardará en el elemento pulsos[1][1], el valor HIGH de la tercera pareja se guardará en el elemento pulsos[2][0], y así. */

```
word pulsos[100][2];
//Índice del pulso que estamos guardando en ese momento en el array byte pulsoactual =
0;
setup(){
        Serial.begin(9600);
}
loop(){
        word contadortiempohigh =0;
        word contadortiempolow   =0;
```

/*La siguiente línea es igual a "  while(digitalRead(irPin)){"  , pero no la escribimos así porque la función digitalRead() es demasiado lenta para poder leer las rápidas variaciones de la señal modulada. Por tanto, hemos recurrido a código C optimizado para la plataforma AVR, el cual nos da un acceso más directo al hardware de la placa. No profundizaremos más en ello.
*/

```
        //Mientras se recibe una señal HIGH
        while ((PIND & _BV(irPin))) {
                contadortiempohigh++;



                //Cuento  unos  cuantos  microsegundos  más
                delayMicroseconds(RESOLUCION);
```

/*Si el pulso es demasiado largo significa que ya se acabó el patrón, por lo que lo imprimimos, nos preparamos para rellenar otra vez el array y, mediante la función return, volvemos al inicio de la función loop()*/

31

```
            if ((contadortiempohigh >= MAXPULSO) && (pulsoactual != 0)) {
                    printpulsos();
                    pulsoactual=0;
                    return;

            }

    }
    /*Ya se ha dejado de recibir un pulso HIGH,
     por lo que guardamos el tiempo que ha durado.*/
    pulsos[pulsoactual][0] = contadortiempohigh;
    //Seguimos leyendo el siguiente pulso de la señal por el pin 2
    //Mientras se recibe una señal LOW
    while        (!(PIND        &        _BV(irPin)))        {
            contadortiempolow++;
            delayMicroseconds(RESOLUCION);
        if ((contadortiempolow >= MAXPULSO) && (pulsoactual != 0)) {
                    printpulsos();
                    pulsoactual=0;
                    return;

            }

    }
    pulsos[pulsoactual][1] = contadortiempolow;
//Hemos leído una pareja de valores HIGH-LOW con éxito. Continuamos pulsoactual++;
}
void printpulsos() {
    byte i;
    Serial.println("OFF \tON");
    for (i = 0; i < pulsoactual; i++) { Serial.print(pulsos[i][0]  *
            RESOLUCION);        Serial.print("        usec,        ");
            Serial.print(pulsos[i][1] * RESOLUCION);
            Serial.println(" usec");

    }
}
```

However, there is a much easier way of modulated signals to use: download and install the library "Arduino IR" available at the https://github.com/shirriff/Arduino-IRremote page. With it we will not have that

acquire any specific new command (so we can recycle any we have at hand a useless device) nor have to investigate your particular pattern so that it can control. This library allows our Arduino can both send and receive remote control codes and multiple patterns including "factory" (supports NEC, Sony SIRC, Philips RC5 and RC6 protocols, and many more) and even has a mechanism to add more protocols if necessary. With it, our Arduino could even function as a forwarder code, receiving a command to trade away and relaying them elsewhere. Consider some examples.

7.11 Example: Assuming you have connected the IR sensor TSOP38238 (or similar) to a digital input pin any of Arduino, the following code displays the "Serial Monitor" commands received from a remote control either.

```
#include <IRremote.h>
/*Pin de entrada digital de la placa Arduino donde hemos colocado la
  patilla de señal del receptor */
int pinreceptor = 11;
//Creo un objeto llamado " irrecv" de tipo IRrecv
IRrecv irrecv(pinreceptor);
//Declaro una variable de un tipo especial, " decode_results" . decode_results
resultados;
void              setup(){
    Serial.begin(9600);
    irrecv.enableIRIn(); //Inicio el receptor
}
void loop() {
/*Miro si se ha detectado algún patrón IR modulado. Si es así, lo leo y lo guardo enteramente
en la variable especial " resultados" , en
forma de número hexadecimal*/
    if (irrecv.decode(&resultados) !=0) {
```

```
/*Primero miro qué tipo de patrón comercial es,

si es que es de alguno reconocido por la librería */

if      (results.decode_type      ==      NEC)      {

                    Serial.print("NEC: ");

}    else    if    (results.decode_type    ==    SONY)    {

                    Serial.print("SONY: ");

}    else    if    (results.decode_type    ==    RC5)    {

                    Serial.print("RC5: ");

}    else    if    (results.decode_type    ==    RC6)    {

                    Serial.print("RC6: ");

   }    else    if    (results.decode_type    ==    UNKNOWN)    {

                    Serial.print("Desconocido: ");
```

```
        }
        /*Y seguidamente muestro el patrón recibido
        (en  formato  hexadecimal)  por  el  canal  serie  */
        Serial.println(resultados.value, HEX);
        /*Una vez el patrón ha sido decodificado, reactivo otra vez la escucha para poder
        detectar el siguiente posible patrón */ irrecv.resume();
    }
  /*Aquí se pueden hacer otras cosas mientras se espera a
    recibir un comando IR*/
  }
```

As expressed in the comments of the previous code, the decode () function is non-blocking; this means that the sketch can perform other operations while waiting the detection of a new pattern.

Each button on a remote control is associated with an identifiable particular pattern (usually from 12 to 32 pulses) by a hexadecimal number. If the button is pressed, this code is usually constantly repeated, although there are controls that only send the code once and use other methods to detect the end of a pulse (and marks the end of pulsation or special codes that work as accountants , etc.).

7.12A example: Once you know the pattern of a remote control, can develop sketches that do react to our Arduino depending on which button has been pressed. For example, the following sketch we can see the "Serial monitor" the down of a remote control of a television, specifically the Sony brand button.

```
#include <IRremote.h>
int pinreceptor = 11;
IRrecv              irrecv(pinreceptor);
decode_results    resultados;    void
setup(){
    Serial.begin(9600);
    irrecv.enableIRIn();
}
void loop(){
    int i;
    if (irrecv.decode(&resultados)!=0)
    {
```

```
accion();

            /*Los mandos Sony envían 3 veces el mismo patrón repetido. Este " for"
            sirve para ignorar el 2° y 3° patrón.*/

            for (i=0; i<2; i++) {
                    //Recibo el siguiente patrón y no hago nada
                    irrecv.resume();

            }

    }
}
void accion() {
    switch(results.value) {
        case 0x37EE: Serial.println("Favoritos"); break;
        case 0xA90: Serial.println("Encendido/Apagado"); break;
        case 0x290: Serial.println("Mute"); break; case 0x10:
        Serial.println("1"); break; case 0x810: Serial.println("2");
        break; case 0x410: Serial.println("3"); break; case 0xC10:
        Serial.println("4"); break; case 0x210: Serial.println("5");
        break; case 0xA10: Serial.println("6"); break; case 0x610:
        Serial.println("7"); break; case 0xE10: Serial.println("8");
        break; case 0x110: Serial.println("9"); break; case 0x910:
        Serial.println("0"); break;
        case 0x490: Serial.println("Aumentar volumen"); break; case 0xC90:
        Serial.println("Disminuir        volumen");        break;        case        0x90:
        Serial.println("Aumentar        canal");        break;        case        0x890:
        Serial.println("Disminuir canal"); break;
        default: Serial.println("Otro botón");
    }
    delay(500);
}
```

.12B Example: To do the opposite, ie, to send a particular trade pattern by an infrared LED connected to our Arduino, it would be as easy as running something like this sketch. Using this sample code, our Arduino on (or off) a television using the Sony protocol whenever you get any data through its serial channel. A very important detail to note is that for this code to work, the infrared LED to be necessarily connected to the digital output pin No. 3 of our Arduino.

```
#include <IRremote.h> IRsend
irsend;
    void setup(){
        Serial.begin(9600);
    }
    void loop() {
        if (Serial.read() != -1) {
        /*El patrón se ha de enviar tres veces porque el protocolo
        Sony lo establece así */
                for (int i = 0; i < 3; i++) {
    /*La función sendSony() envía un patrón especificado como primer parámetro perteneciente a
    ese protocolo (en este caso, el de encendido/apagado). El segundo parámetro indica el
    número de bits que componen ese patrón. Este número es simplemente el número de dígitos
    hexadecimales de los que se compone el patrón multiplicado por 4.*/
                        irsend.sendSony(0xA90, 12);
                        delay(100);
                }
        }
    }
```

There are other predefined different functions sendSony () to send another type of common commercial protocols such as sendNEC (), sendRC5 (), sendRC6 (), sendSharp () or sendPanasonic (). All of them work with the same two parameters. In addition, you can use the sendRaw () function to send a pattern that does not come by default encoded within the library. This function takes three parameters: the first is an array of word values that together form the complete pattern, the second is the size of the array and the third is the modulation frequency in kHz signal to send (that for our projects It will always be

38). The most interesting of this feature is that we can use it to simulate any trade protocol even though this does not come by default encoded within the library. This can be done by choosing from previously sketches of example that come with the library one called "IRrecvDump.ino" and executing our Arduino connected to an IR sensor. By doing this we can see in the "Serial Monitor" precisely the values word to include in the array (the first parameter sendRaw ()), and also, incidentally, the size of the array (the second parameter sendRaw ()).

## TEMPERATURE SENSORS Thermistors

A thermistor is a resistor that changes its

resistance with temperature. Technically, all resistors are resistance thermistors as always changes slightly with temperature, but this change is usually very small and difficult to measure. Thermistors are manufactured so that its resistance changes dramatically, so that can change 100 ohms or more per degree centigrade.

There are two types of thermistors, NTC calls (for "negative temperature coefficient") and PTC (for "positive temperature coefficient"). In the first, as temperature increases, resistance decreases; in the second, as the temperature rises, their resistance increases. In our projects we use normally NTCs to measure temperature; PTCs are typically used more in resettable fuse (where if the temperature increases, increase their resistance to "drown" the current and protect it from over heating circuits).

Thermistors are much cheaper than other types of temperature sensors; They are also waterproof (resistors are only the end of the day) and work at any voltage. They are difficult to damage due to its simplicity and incredibly accurate in measurements. For example, a thermistor 10 k (nominal value, taken at 25 ° C as standard reference) can measure temperature with a margin of error of ± 0.25 ° C (assuming the analog-digital converter is also accurate enough) . However, they not normally withstand temperatures beyond 100 and some degree, and its time constant (ie, seconds needed by the thermistor to reduce by 63% the difference between the initial temperature and final) is usually more ten seconds.

To measure the resistance of a thermistor can be used a multimeter, like any resistance. The value we get depend on the temperature of where we are.

If we want to measure the actual temperature that a plate
Arduino, it must first measure the resistance of the thermistor, and from

it deduce the corresponding temperature. But our Arduino does not have a meter built resistances, so, just as we did with the photoresists, we must use the analog inputs of the plate to detect voltage variations and deduce from these the desired value of resistance current. Thus, the wiring diagram is identical to that already use with photoresists (and photodiodes) must connect a terminal thermistor to food (for example, the 5V pin on the Arduino board) and the other to connect in series terminal fixed resistance value (resistance will "pull-down"); the other terminal of the resistance "pull-down" must be grounded. In addition, we connect an analog input of our Arduino to an intermediate point between the two resistors to get a reading of the potential drop between that point and ground.

By just described, when we detect that the measured voltage is increasing circuit, we can deduce, purely by Ohm's law, the resistance of NTC thermistor is decreasing and therefore the temperature is rising (and vice versa: if the voltage decreases, resistance increases and the temperature also decreases). The exact relationship between the measured voltage and the thermistor resistance can be calculated in the same way that we saw when we studied the photoresists, using the formula $Vavg = (Rpull / (Rpull + Rtermistor)) \cdot Vfuente$.

And we also know that, in reality, is not the voltage Vavg we work with our Arduino, because it always uses an analog-digital converter with performing a "mapping". This mapping converts the read analog value (which can range ranging from 0 V to 5 V assuming that the voltage provided by the source is 5 V) to digital values (ranging from 0 1023). These digital values are what Arduino actually understand and those who work in our sketches. The conversion of analog to digital can be expressed by the same rule of proportionality that we saw with the photoresists: $Vconvertido = Vavg \cdot 1023/5$. From here, as we did with them, if we substitute this expression in the formula above, and cleared away Rtermistor arrive at the following expression: $Rtermistor = (Rpull \cdot 1023 / Vconvertido) - Rpull$, which allows us to know finally what the current value of the thermistor resistance from digitized voltage obtained by the Arduino. The above expression shows, as already knew, which is inversely proportional to Rtermistor Vconvertido.

But what must have pull-up value? The most commonly used value for the "pull-down" resistance that accompanies our thermistor is 10 KQ, but sometimes 1 kW is used. Both values are widely used, but if we want to have more control of the sensitivity of the thermistor would have to

42

choose its value with a little more discretion. To do this, the procedure would replace the same formula already known Vavg = (Rpull / (Rpull + Rtermistor)) · Vfuente different Rtermistor concrete numerical values corresponding to known temperatures (this can be found in the table of equivalence of datasheet) and choose a certain value Rpull: Vavg observing what values we obtained we can choose the pull-up value that allows us to have a wider range without saturating Vavg.

The natural next step, once known the value of the thermistor resistance, find out what temperature would correspond. The easiest way to achieve this is to check the equivalence table always comes in the datasheet, which for certain values of Rtermistor already directly specify the corresponding temperature.

This system will be fine if we just want to design a circuit to make quick comparisons of the type "if the temperature is below this do this and if it exceeds that do that." If we want, however, to obtain the values of real and accurate temperature, we are fortunate to have a mathematical equation that does this with very good approximation (in fact, achieved error of ± 0.02 ° C on a range 100 oC). This is the Steinhart-Hart equation:

ବ = ဖ + ဗ ሺ. ႁ ሻ · ln + ln ⌄ · ሺ. ሺ. ႁ ሻሻ 3 where R is the ohmic value of the thermistor in

ሷ

point, T is the temperature in degrees Kelvin (one Kelvin is
equal to one Celsius + 273.15) and A, B and C are coefficients that are different depending on the type and model of the thermistor (and they are valid only for a given temperature range specified in the datasheet).

However, as this formula is somewhat complex and requires knowledge of the value of various coefficients that may not know, in general we can
using a simplified equation:

ବ = ବ

ሷ ሷ ᙹ

+ ବ · ln ቀ ᏻ○○ᴨ

ᏻ○○ᴨ ᙹ

Φ where T0 is called

nominal temperature (usually 25 ° C = 298.15 K), R0 is the resistance of the thermistor
at that temperature (the so-called "nominal resistance" available data in the datasheet)
and B is the only factor we need to know, which can be found in the datasheet always.

Example 7.13: Once all previous theory, we can implement the test circuit and start writing code
known. As we have said, we will connect a thermistor terminal to 5 V and the other will connect to a
terminal of the "pull-down" resistance. The other terminal of the latter will connect to ground and
finally we use a "third wire" for connecting a center point between the two devices to an analog pin
Arduino (for example, number 0).

The code below shows the channel number Rtermistor values as calculated from the reading
obtained from the analog input. Any temperature calculation is performed.

```
const int Rpull=10000;
void            setup(void)            {
        Serial.begin(9600);
}
void loop(void) {
//El tipo de datos es importante para los cálculos posteriores float lectura;
        lectura = analogRead(0); Serial.print("Lectura analógica
        directa "); Serial.println(lectura);
        /*Convierto " lectura" , que es un valor de voltaje, en resistencia,
          según la fórmula ya vista.*/
        lectura    =    (Rpull    *    1023/lectura)       -    Rpull);
        Serial.print("Resistencia       del       termistor       ");
        Serial.println(lectura);
        delay(1000); //Para la estabilidad en las lecturas
}
```

The following piece of code performs all the necessary calculations for reading in degrees Celsius from the Steinhart-Hart equation simplified. These lines would have to add the previous code, just before the last line delay (1000) ;. To make it work we must also declare and initialize a number of extra variables: "termistorNominal" with value 10000, "temperaturaNominal" with value 25 and "coeficienteB" with the concrete value of our thermistor (in our case, 3950).

```
float steinhart;
steinhart = media / termistorNominal;              // (R/Ro) steinhart = log(steinhart);
// ln(R/Ro) steinhart /= coeficienteB;              // 1/B * ln(R/Ro) steinhart +=
1.0/(temperaturaNominal + 273.15); // + (1/To) steinhart = 1.0 / steinhart;        // Se
invierte steinhart -= 273.15;              // Se convierte a °C Serial.print("Temperatura:
");
Serial.print(steinhart);
Serial.println(" *C");
```

The returned after running the lines above value is a number with two decimal digits. Does this mean that it has an accuracy of 0.01 ° C? Do not! The thermistor has errors and also analog-digital converter. Fortunately, we can

estimate the error of our measured from the nominal thermistor error (searchable data in the datasheet by the name of "tolerance"). Suppose the nominal temperature (25 ° C) have a thermistor 10000 $\Omega$ with an error (tolerance) of 1%, that is, 100 $\Omega$. This means that for that temperature could read values from 9900 to 10100 $\Omega$ $\Omega$. The Steinhart-Hart equation simplified it follows that, if the thermistor has a coefficient B 3950 for example, a difference of 450 $\Omega$ up and down represent 1 ° C difference, so an error of 100 $\Omega$ up and down (tolerance) means an error of about

+0.25 ° C.

Unfortunately, although there thermistors with tolerances even the
0.1% (which can reduce the error to +0.03 ° C), the inaccuracy in the
actually measures is always greater, because there is always at least another source of error that can not forget: the one produced by the analog-digital converter. In the case of this converter in Arduino (which has a resolution of 10 bits) for each bit error measurement can vary up resistance 50 ohms real. This represents an error of +0.1 ° C more to add to the previous errors. In general, therefore, it is advisable to assume a no better than + 0.5 ° C accuracy.

Also keep in mind that the self heating thermistor undergoes a result of the power dissipated, which reaches a temperature above ambient, which is in turn the temperature detected. This forces to limit the voltage value (or current, remember that P = V · I) with which it feeds. In this sense, the "pull-down" resistance is very helpful, because it acts as a voltage divider.

Example 7.14: In general, when analog value readings are made, there are two tips and serving to improve the accuracy of the results. One is to use the pin
Analog 3.3 V as a reference voltage (for increased accuracy of the analog-digital conversion) and the other is to take a set of readings and get the average as the result used (to prevent fluctuations or noise in the measurements).

To achieve the first trick, simply connect the pin "3V3" plate your pin AREF (plus add a line with analogRead () specific to our code). To achieve the second trick, modify the code to include loops to collect various sizes (five are good) and make appropriate means. Thus:

```
/*Dependiendo del número de muestras, la media tarda más
(por el bucle) pero es más suave*/ const int numMuestras=5;
const int Rpull=10000;

int muestras[numMuestras];
void            setup(void)            {
        Serial.begin(9600);
        analogReference(EXTERNAL);
}
void loop(void) {
        byte i;
        float media;
        for (i=0; i< numMuestras; i++) {
            muestras[i] = analogRead(0);
            delay(10);  //Para la estabilidad entre lecturas
        }
        //Calculo la media media
        = 0;
        for (i=0; i< numMuestras; i++) {
            media = media + muestras[i];
        }
        media    =    media/numMuestras;    Serial.print("Lectura
        analógica media: "); Serial.println(media);
        //Convierto " media" a resistencia, según la fórmula media= (Rpull *
        1023/media)    - Rpull); Serial.print("Resistencia  del  termistor ");
        Serial.println(media);
        delay(1000);
}
```

## The analog chip TMP36

This chip uses a solid-state technology to measure the temperature: as the temperature increases, the voltage drop between the base and emitter of a transistor also increases a known quantity. Amplifying this voltage change, an analog signal that is directly proportional to the temperature is generated. Such sensors are accurate, do not wear, do not need calibration, can work under various climatic conditions, are fairly inexpensive and easy to use. In Adafruit it is distributed with product code 165 and SparkFun with code 10988.

Other technical details: its temperature range goes from
-4 0 ° C to 125 ° C and output voltage range

starts from 0.1 V (at -40 ° C) and increases 10 mV per degree Celsius up to 1.75 V (125 ° C). Moreover, its internal circuitry to work, it needs to be fed by a source of between 2.7 V and 5.5 V and 0.05 mA.

Their connections are simple: if you have opposite its flat side (as shown in the figure), the pin on the far left (# 1) is to be connected to food and more to the right (# 3) to Earth. The middle pin (# 2) it is used to obtain a linearly proportional to the temperature (and independent of the voltage provided by the power supply) analog voltage. Therefore, this number pin 2 must be connected to an analog input pin of our Arduino.

Then, to convert the voltage read temperature, simply use the following formula extracted datasheet: $T = (V - 0.5) * 100$ where T would be measured in degrees centigrade and V in volts. For example, if a voltage of 1V is measured, the temperature would be $(1V - 0.5) * 100 = 50$ ° C. From the above formula we can deduce a few things: that the resolution is 10 mV per degree (as already mentioned) and 0 ° C there is a voltage of 500 mV. This feature allows the sensor readings return to freezing temperatures without us having to worry about managing negative voltage values.

If you want to try this chip before adding to our projects, we can use a multimeter to DC voltage measurement mode. However we will have to feed the chip duration of the measurement. To do this, we can use two AA batteries (3 V is well within the voltage range allowed by the chip), so we connect the positive and negative terminals to the corresponding pins. We connect the multimeter to pin 3 (earth) and the pin 2 for measurement. In a room at 25 ° C, the voltage should be about 0.75 V. You can squeeze your fingers play on the encapsulation (to heat) or put you in contact with ice (to cool it) and see changes measured values.

7.15 Example: The following is an Arduino code that displays the channel temperature series (already calculated) environment. The circuit consists of only the component connected to the proper pin-female Arduino (power, ground and analog input pin, which will assume the No. 0) and you're done. The supply voltage may be 5 V as usual or 3.3 V provided by the pin "3V3" the results are independent of that. What we can do is use the button "3V3" also as connecting the reference voltage pin "AREF" and indicating this in the code: with this the accuracy of the results would be improved.

```
void setup(){ Serial.begin(9600); }
void loop() {
        int lectura;
        float voltaje;
        float temperaturaC;
        lectura = analogRead(0);
/*Convierto la lectura (0-1023) en voltaje (0-5). Si el chip se ha alimentado con el pin
" 3V3" , en la fórmula se ha de usar 3.3 en vez de 5.0 */
        voltaje  =  lectura  *  5.0  /  1024.0;
        Serial.print(voltaje); Serial.println(" voltios");
//Convierto este voltaje en temperatura mediante la fórmula temperaturaC =
        (voltaje - 0.5) * 100 ; Serial.print(temperaturaC);
        Serial.println(" grados C");
        delay(1000);  //Para la estabilidad de las lecturas
}
```

49

From here, you can make many interesting projects. For example, we might have a circuit formed by the sensor and three LEDs, one red, one blue and one green. When detected a temperature higher than a preset upper threshold, it may light the red LED, when the temperature is lower than a preset lower threshold, may light when the blue LED and the temperature was between these two thresholds, It could illuminate the green LED. It is left as an exercise.

Example 7.16: A code something more elaborate that the previous example is as follows, in which a circuit is assumed by the presence of, besides the TMP36 sensor, an LCD alphanumeric compatibility with formal LiquidCrystal library Arduino and two buttons ( connected to the digital inputs No. 2 and No. 3 of the Arduino). The display will show the current temperature continuously except when you press a button, at which point it briefly displays the minimum and maximum temperature recorded since the start of the sketch. If you want to delete those records and get a new minimum and maximum temperature from the current, you should press the other button.

```
#include <LiquidCrystal.h>
    LiquidCrystal lcd(7, 8, 9, 10, 11, 12);
    float Tmin = 0; float
    Tmax = 0; void
    setup() {

    //Podríamos haber hecho Tactual global, como Tmin o Tmax float Tactual=0;
      lcd.begin(16,2); lcd.clear();
      //Establezco un mínimo y máximo inicial a partir de la T actual
      Tactual=obtenerTemp();
      Tmin=Tactual; Tmax=Tactual;
      pinMode (2,INPUT); //Botón para redefinir la T máxima y mínima pinMode (3,
      INPUT); //Botón para mostrar la T máxima y mínima
    }
```

```
void loop() {
    float           Tactual=0;
    Tactual=obtenerTemp();
    //Si la temperatura actual supera los límites, los redefino if (Tactual<Tmin) { Tmin
    = Tactual; }
    if (Tactual>Tmax) { Tmax = Tactual; }
    mostrarTempActual(Tactual);  delay(100);
    //Si pulso el botón de "reset"
    if (digitalRead(2) == HIGH) { Tmin = Tactual; Tmax = Tactual; }
    //Si pulso el botón de mostrar máximo y mínimo if
    (digitalRead(3) == HIGH) {
        mostrarTempMin(); delay(3000);  mostrarTempMax(); delay(3000);
    }
}
/*Función que obtiene directamente la temperatura a partir del voltaje leído en
la entrada del sensor TMP36 */
float obtenerTemp(){ return (((analogRead(5)*5.0)/1024)-0.5)*100; }
/*Función que muestra la temperatura actual. Como Tactual es una variable local de loop(), la
tengo que pasar como parámetro. En cambio, con Tmax y Tmin, al ser globales, eso no es
necesario */
void mostrarTempActual(float Tactual){
    lcd.clear(); lcd.setCursor(0,0); lcd.print("Temperatura actual:");
    lcd.setCursor(0,1);  lcd.print(Tactual);  lcd.print(" C/");
}
//Función que muestra la temperatura mínima void
mostrarTempMin(){
    lcd.clear(); lcd.setCursor(0,0); lcd.print("Temperatura mínima:");
    lcd.setCursor(0,1);  lcd.print(Tmin);  lcd.print(" C/");
}
```

```
//Función que muestra la temperatura máxima void
mostrarTempMax(){
    lcd.clear(); lcd.setCursor(0,0); lcd.print("Temperatura máxima:");
    lcd.setCursor(0,1); lcd.print(Tmax); lcd.print(" C/");
}


// We could have gone global Tactual, as Tmin or Tmax float Tactual = 0;
lcd.begin (16.2); lcd.clear ();
// We establish minimum and maximum starting from the current T
ObtenerTemp Tactual = (); Tmin = Tactual; Tmax = Tactual;
pinMode (2, INPUT); // Button to redefine the maximum and minimum T pinMode (3, INPUT); //
Button to display the maximum and minimum T
}
void loop () {
Tactual float = 0; ObtenerTemp Tactual = ();
// If the current temperature exceeds the limits, redefine if (Tactual <Tmin) {Tmin = Tactual; }
if (Tactual> Tmax) Tmax = {Tactual; }
mostrarTempActual (Tactual); delay (100);
// If I press the button "reset"
if (digitalRead (2) == HIGH) {Tmin = Tactual; Tmax = Tactual; }
// If I press the button to display maximum and minimum if (digitalRead (3) == HIGH) {
mostrarTempMin (); delay (3000); mostrarTempMax (); delay (3000);
}
}
/ * Function temperature obtained directly from the voltage read TMP36 sensor input * /
obtenerTemp float () {return (((analogRead (5) * 5.0) / 1024 -0.5) * 100; }
/ * Function that displays the current temperature. As Tactual is a local variable loop (), the I
have to go as a parameter. Instead, with Tmax and Tmin, to be global, it is not necessary * /
mostrarTempActual void (float Tactual) {
lcd.clear (); lcd.setCursor (0,0); lcd.print ("Current temperature");
lcd.setCursor (0.1); lcd.print (Tactual); lcd.print ("C /");
}
```

52

```
// Function that shows the minimum temperature mostrarTempMin void () {
lcd.clear (); lcd.setCursor (0,0); lcd.print ("Minimum temperature");
lcd.setCursor (0.1); lcd.print (Tmin); lcd.print ("C /");
}
// Function to show the maximum temperature mostrarTempMax void () {
lcd.clear (); lcd.setCursor (0,0); lcd.print ("Maximum temperature");
lcd.setCursor (0.1); lcd.print (Tmax); lcd.print ("C /");
```

## The 1-Wire digital chip DS18B20 and protocol

The manufacturer Maxim (formerly Dallas Semiconductor) produces a family of electronic components that can be controlled using a proprietary communication protocol called "1-Wire", which enables you to connect to our Arduino multitude of these components through a single cable data (hence its name). Another remarkable feature is that the components interconnected by this protocol can be located long distances (up even

30 meters).

The DS18B20 digital thermometer is a chip that uses 1-Wire protocol. It is very popular because of its low price and ease of use. It is capable of measuring temperatures in the range -10 ° C to 85 ° C with an accuracy of ± 0.5 ° C. In Sparkfun you can find the product code number 245, in Adafruit with the code number 374 and Freetronics we can acquire breakout shaped plate, called "Temperature sensor module", among others.

If you look straight smooth side of encapsulation, we can see that has three pins: the leftmost corresponds to the ground, the center pin is the digital output data signal (to be connected to an input Arduino digital) and the right pin is used for food, which may well be 5 V provided by the Arduino. Therefore, in principle we need three cables to use this component. It is also very important for the sensor to operate, connect a resistor

4.7 KQ between pin data signal and the supply pin.

However, this chip (as the other components 1-Wire, actually) has the characteristic of power connected differently, using only two wires. This is very convenient in remote sensing distance of our Arduino and / or located in an outdoor environment. It is what is called "parasite", so that the power required by the "parasite" of the data signal. Specifically, the connections are left pin to ground (as before), the center pin to a digital input on the Arduino board (as above) and the right pin must be linked directly to the left pin (for connection to earth ). It is also very important to connect a resistance of 4.7 KQ between digital input pin on the Arduino board where the center pin and 5 V power supply is connected

54

Another interesting feature offered by components 1-Wire is the ability to connect together to form a network (in our case, temperature sensors) using only two cables (ie, exploiting the "parasite" mode), regardless the number of connected components. The wiring in this case is as follows: connecting a first sensor should in parasite mode as described in the preceding paragraph (including resistance 4.7 KΩ), and the following sensors must have all his left pin connected to the left pin of the first sensor (to keep the same common ground), its center pin connected to the center pin of the first sensor (to share the same -and only- data cable) and his right leg also connected to the left pin of the first sensor.

Connecting multiple 1-Wire components to a single cable is possible because each has a unique internal address consists of a 64 bit code divided by 8 bits to identify the component model (DS18B20 has the ID 0x28), 48 bits to identify the particular component individually and 8 more bits (called CRC) used to check for errors in the identification of that component from the other components of the circuit (such as our Arduino).

To program the Arduino DS18B20 through language, we can download and install the library http://www.arduino.cc/playground/Learning/OneWire "One-Wire". Thanks to her, the Arduino can communicate with any device designed to work under the 1-Wire protocol. However, precisely because this library is generic for any device compatible with 1-Wire (and therefore very flexible and versatile), it is relatively difficult to use, because you have to learn some relatively good internal details of the 1-Wire protocol. If we wish to use one or more specific sensors DS18B20 type, we are fortunate to have a much simpler specific library. This library, called "Dallas Temperature Control Library" and available in http://milesburton.com/Dallas_Temperature_Control_Library, you need the library "One-Wire" to operate has been pre-installed, so you need to include both in our sketches. Thanks to it we can read the ID code of each sensor, set the resolution of the measures (9 to 12 bits) and get the value of the temperature of each sensor in a very simple way.

7.17 Example: The following example code, where it is assumed that the pin data from a sensor connected to the digital input pin No. 2 of the Arduino shown:

```
#include <OneWire.h>
#include <DallasTemperature.h>
```

/*Creo un objeto de tipo " OneWire" . Esta instrucción pertenece a la librería One-Wire, por lo que es genérica para cualquier dispositivo compatible con ese protocolo, no solamente el DS18B20. El valor del parámetro indica el número de pin digital de la placa Arduino donde está conectada la patilla de datos del sensor (o sensores). */

```
OneWire dispositivoOneWire(2);
```

/*Pasamos la referencia del objeto recién creado a la librería

DallasTemperature                */                DallasTemperature

```
sensores(&dispositivoOneWire); setup(){
        Serial.begin(9600);
```

/*Se inicializa la librería. Opcionalmente, se puede especificar un parámetro para ajustar la precisión de las medidas: por defecto es de

9 bits, pero puede aumentar hasta 12. La contrapartida es una mayor lentitud en las lecturas*/

```
        sensores.begin();
}
loop(){
```

/*Solicito las temperaturas de todos los

posibles     sensores     presentes     en     el     canal     de     datos     */

```
        sensores.requestTemperatures();    Serial.print("Temperatura    para    el
        dispositivo 1 es: ");
```

/*El parámetro de getTempCByIndex indica el número de sensor del cual se quiere obtener la temperatura en grados Celsius: el " 0" se corresponde al primer sensor presente en el cable de señal de datos, el " 1" sería el siguiente, y así*/

```
        Serial.print(sensores.getTempCByIndex(0));
}
```

The insert breakout TMP421

Modern Device distributes an insert which includes analog chip TMP421. We must connect the analog pins No. 2, 3, 4 and 5 of the Arduino board. The first will feed plate, the second to connect to ground, and the last two are used to set the I2C channel through which communication between the two devices is performed. It can measure temperatures from -40 oC to 125 oC with an error of +1 oC. It is really easy to use by library available https://github.com/moderndevice/LibTempTMP421.

## Humidity sensors Sensor DHT22 / RHT03

This section specifically discuss the digital sensor

temperature and humidity RHT03 of Maxdetect (http://www.humiditycn.com), which (along with many others) distributed Sparkfun with product number 10167 and Adafruit (here with the name and number DHT22 product No. 385). You can also find it by the name of AM2302. This sensor is very basic and slow (you can only retrieve data at least once every 2 seconds), but it is low cost and very manageable to obtain basic data on home projects.

His most notable technical features are: can be fed with a voltage of 3 V and 5 V and

2.5 mA maximum measurable range

temperatures between -40 and 125 ° C with an accuracy of ± 0.5 ° C and a humidity range between 0 and 100% with an accuracy of 2-5%. It is basically formed by a capacitive humidity sensor and a thermistor.

This chip has four pins; facing him are the supply (# 1, the more the

left), the digital output data (# 2), one not connected to anything and that

You can ignore (No. 3) and the ground (No. 4, the rightmost). So, for our Arduino to read the data issues this chip, we connect your pin # 1 to pin-socket plate 5 V (for example), their No. pin 4 to pin "GND" and pin # 2 to a pin-socket digital input. In addition, you should connect a resistor ("pull-up") of 4.7 KQ between No. 1 and No. 2 pin.

Unfortunately, the protocol used by the sensor to transmit the digital data is not standardized, so in principle should learn their inner workings to correctly interpret the information that we are coming at a certain time. Fortunately, we have at our disposal many bookstores Arduino compatible with this sensor. These libraries, despite being different, are largely equivalent, because all they do basically is to allow focus on simple data extraction temperature and humidity without having to learn the specifics of the particular protocol used by the chip. Therefore, the choice of a library or another is not critical.

Example 7.18: The people of Adafruit, for example, has scheduled a compatible Arduino library with this sensor (https://github.com/adafruit/DHT-sensor-library) Once installed like any other library, we can try running the following skit example, which shows the "Serial Monitor" temperature and humidity. To see changes in the data, you can try to expel breath on the sensor if you want to increase the humidity read:

```
#include <DHT.h>
/*El primer parámetro es el número del pin-hembra de la placa Arduino donde está
conectado el pin de datos del sensor. El segundo dato especifica el modelo de sensor
que se está utilizando (ya que la librería puede servir para varios. El valor
correspondiente para el DHT22 es " 22" . Para consultar los otros valores posibles, se
puede mirar el archivo " DHT.h"  que viene incluido dentro de la librería.*/ DHT dht(2, 22);
void setup() { Serial.begin(9600);
        dht.begin();
}
void loop() {
        float h,t;
        //Las lecturas del sensor pueden tardar hasta 2 segundos h =
        dht.readHumidity();
        t = dht.readTemperature();
/*Compruebo que los datos recibidos son válidos. Concretamente, la función isnan() mira si el
dato no es un número (" IS Not A Number" ), en cuyo caso algo ha ido mal*/
        if (isnan(t) || isnan(h)) { Serial.println("Algo falló ");
        } else {
```

```
                Serial.print("Humedad:            ");
                Serial.print(h); Serial.print(" %\t");
                Serial.print("Temperatura:            ");
                Serial.print(t);
                Serial.println(" *C");
        }
}
```

Example 7.19: If what you want is (instead of displaying the data obtained by the serial channel) save it to an SD card so that they can look calmly

then we should modify the previous sketch seems to be the next. In him we have added extra checks to account for any errors that may occur in reading in the card:

```
#include <SD.h>
#include      <DHT.h>
DHT dht(2,22);
const int intervalo = 10*1000; //Intervalo entre lecturas, en ms long tiempoUltimaLectura = 0;
//Tiempo de la última lectura, en ms long tiempoActual = 0;        //Tiempo actual devuelto
por millis() float h,t;              //Humedad y temperatura obtenidas
void setup() {
    /*Utilizo una función propia para inicializar la tarjeta SD. Si la tarjeta se inicializa
    correctamente, inicializo la librería DHT */ if (inicioSD() == true) {
        dht.begin();
    }
}
```

```
void loop(){
    /*Uso el truco de repetir código cada determinado tiempo (especificado por
    "intervalo") sin utilizar delay() */ tiempoActual = millis();
    if (tiempoActual > tiempoUltimaLectura + intervalo) {
        //Obtengo los datos de temperatura y humedad h =
        dht.readHumidity();
        t = dht.readTemperature();
        //Si las lecturas NO son erróneas (atención al "!")
        if (!isnan(t) || !isnan(h)) {
            //Abro el fichero y escribo a partir de la última línea
            File mifichero = SD.open("datalog.csv", FILE_WRITE);
            if      (mifichero==true)      {      mifichero.print(h);
                mifichero.print("\t");              //Tabulador
                mifichero.println(t); mifichero.close();
                tiempoUltimaLectura = millis();
            }
        }
    }
}
boolean inicioSD() {
    boolean resultado = false;
    //Aunque no se utilice, el pin 10 se ha de configurar como salida pinMode(10, OUTPUT);
    //Suponemos que la tarjeta está conectada al canal SS vía pin 4


    if (SD.begin(4)==false) {
        //Si no lo está, el sketch no hará nada resultado =
        false;
    }
```

/\*Si sí, abro el fichero "datalog.csv" para escribir una línea (lo que sería el título) a continuación de las posibles líneas anteriores que existan de otras ejecuciones previas.\*/

```
else {
    File mifichero = SD.open("datalog.csv", FILE_WRITE);
    if (mifichero == true) {
        mifichero.println();
        mifichero.println("rH (%) \t temp. (*C)");
        mifichero.close();
        resultado = true;
    }
}
return resultado;
}
```

The interesting thing about the previous two sketches is that both send (one serial channel and the other to the SD card) the temperature and humidity data in tabular form, with even a title for the columns. This fact can use to easily import this information into data processing programs (such as spreadsheets), to perform statistical calculations or graphical representation, among other possibilities.

In the case of the file saved on the SD card, to perform this import only thing we do is put the card into a reader connected to our computer and open that file with our favorite program, such as the LibreOffice Calc (software http: //www.libreoffice.org, free spreadsheet application, multiplatform and free calculation) or, if only the graphics drawing is required and nothing more, the LiveGraph (http://www.live-graph.org software, plotter also free, multiplatform and free).

In the case of obtaining the data using the "Serial Monitor" what we do is the following: press within it the key combination CTRL + A to select all the data and press CTRL + C to copy in memory. From here, we can open a new plain text file and press CTRL + V to paste the data, or open a new spreadsheet using for example LibreOffice Calc. In the latter case, once already in leaf calculation, also press CTRL + V to paste the data and then we would see a popup box appears

Configuration import text. In this picture we should choose "tab separated" (or similar) for the data to be properly incorporated.

If we used a different program "Serial Monitor" to see the data received by the serial channel, we could transfer these to a spreadsheet in a more simple way, since the vast majority of them have the option of saving to real time received data in a text file ready to be imported.

Another way to process the data read by the serial channel to allow further analysis of them is to use a free program called MegunoLink (http://www.blueleafsoftware.com). This program can (among other features) to draw on screen and in real time a graphic corresponding to the data received by the serial channel. To work, we use a library that provide us with the same developers MegunoLink, the "Arduino graphing library" in our sketches. Bookseller along with several code examples where you can see their behavior are offered; there you can see how we can configure the names of the X and Y axes of the graph and its various legends. Unfortunately, MegunoLink only works on Windows systems. Another plotter in real time (which also works only for Windows) is SerialChart (http://code.google.com/p/serialchart).

62

Returning to the humidity sensors, other library supports DHT22 / RHT03 sensor is to be found in https://github.com/ringerc/Arduino- DHT22. Another is the http://arduino.cc/playground/Main/DHTLib available. On the other hand, Freetronics distributes a breakout plate with the same sensor DHT22 / RHT03 (under the name "Humidity and Temperature Sensor Module"), which

It has three connectors: land, power and data signal (listed from left to right) and can be programmed using a library of Freetronics own, available in https://github.com/freetronics/DHT-sensor-library, very similar to Adafruit.

## The SHT15 and SHT21 sensors

Another humidity sensor (and temperature) is the SHT15, distributed by Sparkfun breakout shaped plate with product code 8257. This plate consists of four pins: supply (5 V), land, connector "SCK" and connector "DATA" . The latter two are to be connected to two of our Arduino digital inputs. Although the communication protocol using this chip requires the use of two cables as happens with I2C not be confused because they are

different communication systems. As highlight technical data, we can say that reaches an accuracy of up to +0.3 ° C in temperature measurements and +2% in moisture measurements and has a response time of less than 4 seconds.

The easiest way to program this sensor is using available https://github.com/practicalarduino/SHT1x bookstore (also valid for other sensors of the same family manufactured by Sensirion, SHT10 as the STH11, the SHT71 or SHT75).

On the other hand, a chip of the same manufacturer that it is able to use the I2C communication is the SHT21 sensor. LoveElectronics distributes a breakout plate that offers very comfortable four connectors required: power (3.3 V, attention), land, SDA pin and SCL pin. It can be programmed by the library available in https://github.com/misenso/SHT2x-Arduino-Library (also valid for SHT25). Another very similar plate distributes Modern Device, along with the library "LibHumidity", downloadable from the product website.

## The distance sensors sensor Ping)))

The digital distance sensor Ping))) ™ Parallax is capable of measuring distances between about 3 cm and 3 m. It does this by sending an ultrasound (ie, sound frequency too high to be heard by the human ear) through a transducer (one cylinder that can be seen in Figure
lateral) and waits for this ultrasound on an object and bounce back, return that is detected by the other transducer. The sensor returns the time between sending and receiving ultrasound later time. As the propagation velocity of any (ultra) sound in a medium such as air is of known value (consider that is 340 m / s or what is the same, 0.034 cm / μ s-, although this is only an approximation ) passed this time we can use to determine the distance between the sensor and the object that caused the bounce.

The insert in the next shows three pins marked. Given visible across the transducers are, from left to right: the ground pin, that of

supply (5 V) and pin to communicate with a digital pin-female of our Arduino (we'll call it from now "signal pin").

This sensor measures distances only when prompted. To do this, our Arduino sends through its digital pin (connected to signal pin insert) A HIGH pulse of a precise duration of 5 microseconds. This is the signal that triggers the sending of ultrasound. After a short time, the sensor will receive the rebound of ultrasound and as a result, our Arduino receive that data for the same digital pin used previously. At that time, the Arduino can calculate the time between sending and receiving the ultrasound time. The described operation requires that due toggle mode digital Arduino pin connected to the sensor, so that either INPUT or OUTPUT type as appropriate.

64

To calculate the distance, we must remember the formula v = d / t (which is nothing but the definition of velocity: distance traveled in a given time). If we finally obtain d = v · t, where "v" know and believe constant (0.034 cm / ms) and "t" of the above formula cleared the "d" (the distance) is the time measured by the sensor . Keep in mind, however, that the data obtained from the sensor is the total time it takes for the ultrasound to "go and return", so usually want this previously split between two value to assign to "t".

Example 7.20: The following sketch done all these calculations and displays the results via the serial channel. We have assumed that the signal pin insert is connected to digital pin-female # 8 of Arduino. You can try real time putting an obstacle in front of the sensor and moving it back and forth.

```
int distancia;
unsigned long tiempo=0;
void setup(){ Serial.begin(9600);
}
void loop(){
        //Obtengo la lectura de la distancia medida por el sensor enviarYRecibir();
        /*Convertimos el tiempo a distancia, sabiendo
        la velocidad del ultrasonido (en cm/microsegundos) */ distancia =
         int(0.034*tiempo); Serial.print("Distancia: ");
         Serial.print(distancia);
```

```
                Serial.println(" cm");

                delay(500);

}

void enviarYRecibir(){

/*Configuramos  el  pin  de  datos  como  salida  y
estabilizamos el sensor para poder enviarle el pulso
de activación*/

                pinMode(8,        OUTPUT);

                digitalWrite(8,        LOW);

                delayMicroseconds(5);

/*Ahora es cuando enviamos el pulso
de 5 microsegundos para activar el sensor y enviar el
ultrasonido */

                digitalWrite(8,        HIGH);

                delayMicroseconds(5);

                digitalWrite(8, LOW);

/*Mientras el ultrasonido viaja, cambiamos el modo del
pin de datos a entrada
para leer el rebote del pulso */

                pinMode(8, INPUT);
```

/*Medimos la longitud del pulso entrante. El truco aquí está en saber que justo después de enviar el ultrasonido, el sensor comienza siempre a emitir una señal HIGH, la cual será recibida por el pin de la placa Arduino acabado de configurar como entrada (el nº 8 en este caso). Pero en el momento que se reciba el rebote, el sensor automáticamente cambiará esa señal HIGH a LOW. Esto permite utilizar la función pulseIn() para medir el tiempo transcurrido entre un instante y otro: tal como está escrita en el sketch, esta función cuenta el tiempo que pasa desde que el pin nº 8 empieza recibiendo una señal de valor HIGH (en el momento de enviar el pulso) hasta que lo deja de recibir (en el momento de recibir el rebote). La duración de esa señal HIGH recibida se corresponde con el tiempo que queremos medir*/

```
        tiempo=pulseIn(8, HIGH);
```

/*Divido la longitud del pulso a la mitad, porque solo quiero utilizar el tiempo tardado en la ida del ultrasonido, no el

de ida y vuelta*/

```
        tiempo=tiempo/2;
}
```

From the detection of the presence of an object circuit reactions can be diverse: they may include activation or lighting acoustic alarms, or opening or closing different hatches (via servomotors) or any nothing but imagination suggests. For instance,

exercise as possible propose the construction of a radar home: one would have to place the sensor on a servomotor that were continuously moving between 0 and 180 degrees (and vice versa). At the moment an object to a distance less than the set threshold is detected as, it could illuminate an LED or activate a buzzer, for example.

The sensor SRF05

Other digital sensor that uses the method of counting the time between the emission of an ultrasonic pulse and subsequent reception for measuring distances of time is the Devantech SRF05. It is capable of measuring distances from 3 cm to 3 m to a rate of 20 times per second.

This sensor can be connected in two different ways our Arduino: either using four cables ("mode 1" compatible with its predecessor, the SRF04 sensor), or by using three ("mode 2"). Looking at the back of the sensor and maintain the silkscreen "SRF05" visible on the left, in mode 1 the five connectors on the bottom correspond, from left to right, with: supply (5 V) input ultrasonic rebound ( pin "echo") ultrasonic pulse output (pin "trigger"), pin that is not connected to anything and earth. The pin "echo" be connected to a digital input pin of our Arduino pin and "trigger" to a digital output pin. This pin "trigger" is responsible for generating a pulse with a duration of 10 microseconds exact, which marks the start sending the ultrasonic signal, and pin "echo" uses the same trick that the Ping))) sensor to counting the time between sending and receiving ultrasound time: HIGH signal while maintaining a rebound is not received.

If we use the mode 2, the same connectors are (from left to right as well) with: supply (5 V) pin that is not connected to anything, input and output in one of the ultrasonic signal, ground and ground again. In this mode, the sensor uses a single pin to send and receive the pulse rebound (as indeed occurs with sensor Ping)))). This enables a simpler wiring, although programming is complicated because the same pin must be switch between input and output depending on the circumstances.

68

7.21 Example: The following code, which uses mode 1, graduated brightness LED giving a PWM signal will vary depending how close he is an obstacle to the sensor: the idea is that the closer you are, the brighter the LED. Specifically, every centimeter (below 255 cm) nearest the LED brightness is increased at a point of the PWM signal. If you wish to use the maximum range

that enables sensor distances, the LED brightness should be calculated otherwise (eg depending on the percentage of the measured distance to the maximum distance possible), but as presented here, the code is easy.

Another detail worthy code comment is that, to avoid getting noise (ie, highly volatile individual values) and therefore getting smoothed readings, we used a system already seen on other occasions: the calculation of the average last measurements taken.

You may also notice that for the distance, the time between the signal "trigger" and "echo" (that is, in short, which measures the sensor, measured in microseconds) is multiplied by 0.017. Why? Because as explained when the Ping sensor was))), has been made serve the basic formula d = v * t where v = 0.034 (the speed of ultrasound in the air in units of cm / ms, or that is, 340 m / s) and "t" is to be divided by 2 to count only the path "back" and not "return".

```
const int nLecturas=10; //Número de lecturas para hacer media
int lecturas[nLecturas]; //Array que guarda las últimas lecturas int indice = 0;
//Posición actual dentro del array
int total = 0;        //Suma de las lecturas guardadas en el array int media = 0;        //Media de
las lecturas guardadas en el array unsigned long tiempo = 0;
unsigned long distancia = 0;
void setup() {
```

```
pinMode(9, OUTPUT); //Donde está conectado el LED (señal PWM) pinMode(2, INPUT);
//Donde está conectado el pin " echo"  del sensor pinMode(3, OUTPUT); //Donde está
conectado su pin " trigger"
//Creo un array inicialmente vacío
for (int i = 0; i < nLecturas; i++) {lecturas[nLecturas] = 0;} Serial.begin(9600);
}
void loop() {
//Estabilizamos el sensor antes de enviar el pulso de activación digitalWrite(3, LOW);
delayMicroseconds(5);
digitalWrite(3, HIGH); //Envío el pulso de activación delayMicroseconds(10); //La duración
de este pulso ha de ser 10 µs
//Paro la activación y comienza el envío de la señal ultrasónica digitalWrite(3, LOW);
/*Inmediatamente después de empezar el envío del ultrasonido, por
```

el pin " echo"  se empieza a recibir una señal HIGH. La función pulseIn() pausa entonces
el sketch para contar el tiempo transcurrido hasta recibir el rebote, momento en el cual por
el pin " echo"  pasa a detectarse una señal LOW y pulseIn() devuelve su resultado.*/

```
tiempo = pulseIn(2, HIGH);
```

/*Calculo la distancia (distancia en cm = velocidad en cm/µs multiplicado por el tiempo en
µs). Ya se ha dividido entre dos para contar solo el tiempo de ida*/

```
distancia = 0.017*tiempo;
```

/*Pero no quiero obtener un valor de distancia individual, porque puede ser muy
fluctuante: quiero que el valor tenido en cuenta sea una media de los últimos diez valores
individuales medidos. Para ello, primero elimino de la suma total el valor ubicado en el
índice actual del array, que corresponde a la medida tomada hace diez iteraciones */

```
total= total - lecturas[indice];
```

/*Ahora añado la nueva distancia medida precisamente en ese elemento del array,
sobrescribiéndolo. De esta manera, se hace una reescritura cíclica de cada elemento
cada diez lecturas*/ lecturas[indice] = distancia;

```
//Y finalmente, añado este nuevo valor a la suma total otra vez total= total +
lecturas[indice];
indice = indice + 1; //Voy al siguiente elemento del array
//Al final del array (10 elementos) vuelvo a empezar if (indice >=
nLecturas) { índice = 0; }
media = total / nLecturas;  //Calculo la media
/*Si la distancia es menor que 255 cm cambio el brillo del LED, de forma que a menor
distancia más brillante sea */
if (media < 255) {
    analogWrite(9, 255 - media);
}
delay(100); //El tiempo mínimo entre lecturas ha de ser de 20µs
}
```

Si quisiéramos utilizar el modo 2, simplemente tendríamos que haber sustituido las primeras líneas de la función " loop()"  del sketch anterior (hasta la línea de *pulseIn(),* esta incluida) por las siguientes (donde suponemos que el pin " trigger- echo"  del sensor está conectado al pin nº 2 de la placa Arduino):

```
/*Mandamos un pulso bajo de 5 microsegundos para asegurar que siempre se inicie en
LOW, y seguidamente mandamos una señal HIGH que sirve para iniciar las mediciones*/
pinMode(2, OUTPUT);
```

```
digitalWrite(2,          LOW);
delayMicroseconds(5);
digitalWrite(2,          HIGH);
delayMicroseconds(10);
digitalWrite(2, LOW);
/*Como utilizamos el mismo pin para recibir el eco, lo cambiamos de salida a entrada*/
pinMode(2, INPUT);
tiempo = pulseIn(2, HIGH);
```

The sensor HC-SR04

This sensor is similar to the above. It has four pins: "VCC" (it must be connected to a source of 5 V), "Trig" (responsible for sending the ultrasonic pulse, therefore, must be connected to a pin digital output of the Arduino) "Echo" (responsible for receiving the echo of that pulse, then must be connected to a digital input pin of the Arduino) and "GND" (ground). It can be purchased in IteadStudio or ElecFreaks for less than ten euros.

Like the previous sensors, has a range of sensitive distances from 3 cm to 3 m with an accuracy of 3 mm, and its operation is also very similar: after issuing the pin "trigger" signal 10 microseconds to start shipping ultrasonic signal, waiting to detect the echo by detecting the end of the HIGH signal received by the pin "echo".

In fact, the sample code shown for mode 2 SRF05 sensor can be used without any change in this sensor. However, if desired, you can use the "New-Ping", downloadable from http://code.google.com/p/arduino-new-ping library, which provides a simple and common way to manage different models ultrasonic distance sensors such as the HC-SR04 itself (but also the sensor Ping))) and SRF05, etc.).

The sensor LV-EZ0

Another distance sensor using ultrasound is the LV-EZ0 of Maxbotix sensor. However, unlike the above, the LV-EZO is an analog sensor. Therefore, for use with our Arduino we connect (besides the pin "+5 V" to the 5V supply provided by the Arduino board and the pin "GND" to the common ground) the pin labeled "AN" a analog input of our Arduino.

The range of distances that can measure this sensor depends greatly on the size of the obstacle: if this is the size of a finger, the range is about 2.5 meters; if this is the size of a sheet of paper, the range can be increased to 6 meters. In any case, it is not able to detect smaller distances of 30 cm. The good news is that the behavior of this sensor is linear: if an obstacle is for example 2 meters, reading of voltage received by the analog input pin is half that if it is 4 meters. This allows the readings are very easy to perform.

Example 7.22: We can use a simple sketch like this to observe the different distances from an obstacle detected over time.

```
int sensorPin = 0; //El sensor está conectado al pin analógico nº 0 void setup(){
    Serial.begin(9600);
}
void loop(){ Serial.println(analogRead(sensorPin));
    /*El delay() ralentiza la toma de lecturas para hacerlas más fáciles de leer. No
obstante, si queremos detectar la presencia de obstáculos que rápidamente atraviesan de un
lado al otro el espacio sensible, este delay() se podría reducir o quitar */
    delay(100);
}
```

How could modify the above code for adding a LED connected to a digital output pin of our Arduino, this is lit when the presence of an obstacle is detected closer than a certain threshold distance? It is left as an exercise.

There are several versions of this sensor with different ranges and different widths detectable distance sensitive space within which we can recognize obstacles. The EZ1 sensor has a narrower space than EZ0 sensitive, the EZ2 has narrower than the EZ1, and so on EZ4. A narrow space sensitive is better if only you want to detect objects located directly in front of the sensor, and one wide is better if you want to detect any nearby object. Maxbotix also offers a line of more accurate sensors (the "XL") with up to an accuracy of 1 cm, more distance range and better noise suppression (ie, their readings are less shaky). Sparkfun EZ0 distributes sensor with product code 8502, Adafruit with code 979. If you want to find other sensors manufactured by Maxbotix online stores of these dealers, just type "Maxbotix" in the search.

The sensors GP2Yxxx

There is a family of analogue distance sensors manufactured by Sharp and which begin with GP2Y that, unlike those described above, using infrared waves. Generic operation is as follows: the chip contains in its front IR emitter and receiver. If the infrared beam (which is modulated) hits an obstacle, it will disperse in various directions, but also in the direction where the receiver is located exactly. Using the triangulations calculating, from the measurement of the angle of incidence of the infrared beam on the receiver, one can deduce the distance to the obstacle. The output of these sensors is proportional to the distance calculated voltage.

This method has some drawbacks. For example: sunlight (which contains infrared), both direct and reflected in large amounts can disrupt reading. To minimize this problem, it is good practice not to use individual readings but take an average of a number of values as the valid reading. Another disadvantage to be taken into account is the minimum distance from our sensor: There is a maximum angle beyond which the infrared sensors no longer receive the reflected beam successfully, so the measured distances under the minimum values have little meaning.

We choose to study specifically the GP2Y0A02 sensor. This sensor we can find for example in Pololu with product code 1137. Its measuring range is between 20 and 150 centimeters and incorporates one of JST connector 3-pin type, so to plug it into a breadboard need a cable as Product No. 117 or No. Littlebird Electronics Sparkfun 8733, which consists on the one hand a JST 3-pin socket and the other three are free cables without terminal. Each pin of the sensor corresponds to the supply (5 V), land and analog output (to connect to a pin-female analog input on the Arduino board).

This sensor returns a voltage which is inversely proportional to the distance: The greater the distance to the object, the chip generates less voltage (although the relationship is not linear). This fact can use to write composed of a set of "ifs" style sketches "if the measured voltage is within a specified range of values and if something happens in another, something else to happen" if the only thing that interests us It is to compare different distances without having to know its exact value.

If we want, however, it is to know the actual distance of the obstacle,

We can refer to the sensor datasheet precise figure provided by this voltage as a function of the measured distance, so that from the

values shown in this chart could interpret voltage values received and displayed as distance.

Example 7.23: The following sketch, however, uses a formula to obtain the distance. This formula gets precisely the data of the graph of the datasheet, but not exact, so that must be taken with caution. You may need to calibrate numerical data involved (in our case, 65 and -1.1) To finish profiling measurements in our particular case.

```
    float sensor = 0.0;
float   sensor   =   0.0;   float
distancia = 0.0; void setup(){
        Serial.begin(9600);
}
void loop(){
        sensor = analogRead(0);
/*Convierto el dato obtenido por el conversor analógico-digital en un valor   de   voltaje.   Si   el
sensor   se   alimenta   con   3,3V,   hay   que sustituir el 5.0 por un 3.3 */
        sensor = sensor * 5.0 / 1024.0;
//Calculo la distancia a partir del voltaje obtenido distancia =
        65*pow(sensor, -1.10); Serial.print(distancia);
        delay(250);
}
```

A very curious usefulness of distance sensors is to build homemade theremin. A theremin is a musical instrument having a pair of metallic antennas that control the volume and frequency of the emitted note, as we approach or move away from them. Instead of two metal antennas use a distance sensor; G2PY0A02 specifically (but obviously we could have also implemented other models, such as ultrasonic). The idea is to get an interpretable value as distance from the read voltage value of the sensor, and from there generate the corresponding sound.

7.24 Example: A concrete implementation of theremin is the next sketch, which reproduces musical notes different standards depending on the detected distance (the closer you are to the obstacle-our hand, the sharper will be the note). The sound wave is emitted through a buzzer connected to the digital output pin No. 8 of our Arduino. That sound has to last a short enough time (we put 125 milliseconds but can be changed) to
to track real-time motion detection obstacle.

```
int lecturaSensor, nota;
//Array que guarda las frecuencias a reproducir float
frecuencias[] = {
    329.63, // Mi
    349.23, // Fa
    369.99, // Fa#
    392.00, // Sol
    415.30, // Sol#
    440.00, // La
    466.17, // La#
    493.83, // Si
    523.25, // Do
    554.36, // Do#
    587.33, // Re
    622.25, // Re#
    659.26, // Mi
};
byte numFrecuencias=13; //Número de elementos del array void setup(){
        Serial.begin(9600);
        pinMode(8,OUTPUT); //Donde está conectado el zumbador
}
void loop() {
/*Cuanto más lejos esté el obstáculo, " lecturaSensor"  menos valdrá
De todas formas, hay que saber que esta relación no es lineal*/
        lecturaSensor=analogRead(0); //Donde está conectado el sensor
/*Asocio " lecturaSensor"  a una de las frecuencias predefinidas*/
        nota=map(lecturaSensor,0,1023,1,numFrecuencias);
/*Hago que a medida que el obstáculo se acerca, la nota es  más aguda*/
        tone(8,frecuencias[nota]);
/*Tiempo de reproducción mínimo de la nota */
        delay(125);
}
```

The other sensors are similar to GP2Yxxx GP2Y0A02: the biggest difference between them is the range of distances that can be measured and the voltage level provided by this, but its performance is very similar. Both Adafruit Sparkfun and we find for example the GP2Y0A21 sensor (with code # 164 and # 242, respectively) that can measure distances between 10 and 80 cm.

The sensor IS471F

This sensor is not a distance detector itself but simply presence; specifically detects the presence or absence of an obstacle between 1 cm and 15 cm. It works by emitting an infrared beam and seeing if you get bounced. If so, the sensor will generate a LOW signal (which can be read by an Arduino board conveniently connected to it) and if no object is detected, the sensor will generate a HIGH signal.

The sensor consists of four pins, which are (if we look at its flat face from the front, from left to right): supply (5 V) data detection (to be connected to a digital input pin of our Arduino), earth and emission of infrared signal (we'll call this pin "X"). It is recommended to connect a capacitor (0.33 uF) type "bypass" between the pin "X" and the ground pin.

What is more surprising is that this sensor is not able to issue itself no infrared signal, so to function must be connected to pin "X" sensor cathode external infrared LED (preferably 940 nm). The emission of this LED is suitably modulated by a 38 KHz internal modulator containing the sensor, so this is relatively immune to interference from other light as the sun (the rebound is in turn demodulated by an internal demodulator). The anode of the infrared LED should be connected to the appropriate power source, usually through a voltage divider. If the voltage divider turn it into a potentiometer, we can regulate within limits the distance that can detect the object, because the lower the resistance of the more intense potentiometer is the light emitted by the LED and, therefore, the greater the distance.

The QRD1114 and sensors QRE1113

QRD1114 sensor (product code 246 SparkFun) is really nothing more than an infrared emitter and a phototransistor under the same package, so the principle is similar to the analog infrared sensors already seen: the more distance has the obstacle, unless we obtain the output voltage sensor. Its most outstanding feature is its range of distances, since it is only able to detect objects between 0 and 3 cm of him. Actually, this component is not designed to measure exact distances, but only to check the proximity of these objects below 3 cm.

The Arduino code to use with this sensor is identical to that shown with the LV-EZO sensor: it is not only to obtain the values of an analog input. However, the connections are more complex: this sensor has four

pin because actually we have said that simply consists of an infrared LED (anode and cathode) and a phototransistor (collector and emitter). Corresponding to the anode pin it is connected through a voltage divider 200

$\Omega$ to pin 5 V supply our Arduino, the corresponding pin to the collector must be connected to an analog input pin of our Arduino and the other two pins are to be connected to the common ground. In addition, between the anode and the collector must connect a resistor type "pull-up" whose value can range between 4.7 and 5.6 K$\Omega$ K$\Omega$ (as it is, the values read by Arduino change) .

This sensor can also be further used to detect white and black surfaces (hence could be used in building robots followers lines) as white surfaces reflect more light than black, thus obtaining higher readings. However, there is a specific task for this particular sensor: the QRE1113.

The QRE1113 sensor is marketed as a breakout plate Sparkfun (product # 9453), which simply provides three connectors: "VCC" (a plug to a power of 5 V), "GND" and "OUT" (plug in an analog input of our Arduino). In fact, this product is the analog version of the breakout board, as you can also find the digital version. In the analogue version, higher values for the read voltage means that the light emitted by the LED has been reflected most strongly and thus it was better received by the phototransistor (ie, has been detected a light surface).

79

# TILT SENSOR

Tilt sensors are small, inexpensive, and easy to use. They can work with voltages up to 24 V and currents of 5 mA. Consist of a cavity and a conductive inside free mass (such as a metal ball rolling); one end of the cavity has two conductive poles so that when the sensor is oriented with the tip down, the wheel mass to the poles and closes. Therefore, these sensors act as switches, leaving or not passing current circuit according to the inclination.

Although not as precise or flexible as a complete accelerometer, they can easily detect orientation or movement.

Check tilt sensor is easy: put the meter in continuity mode and connect a cable to any multimeter (as tilt sensors are not polarized devices) to each terminal of the sensor. It must then be tilted to determine the angle at which the switch opens and closes.

If this sensor is connected in series to an LED (mandatory and voltage divider) and the circuit is powered, we will see how the LED is turned on or off according to the inclination to give you the design, as if we were using a pushbutton " invisible ".

If we use with our Arduino, we must follow the same recipe that we use when we saw the buttons: can be connected using a resistance "pull-up" or "pull-down" (values between 200 Ω and 1 kW are fine).

Example 7.25: The following circuit shows a basic example where you can see that reading made by the sensor is received in a digital input pin of the plate (we used # 2) and is used to control a LED connected to digital output pin No. 4 (through its voltage divider), which is on or not according to the (HIGH or LOW) value detected by the sensor.

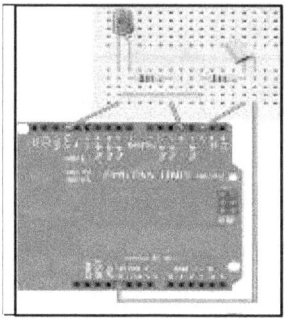

And the code:

```
void                    setup(){
    pinMode(4,OUTPUT);
    pinMode(2, INPUT);
}
void loop(){
    digitalWrite(4,digitalRead(2));
}
```

The previous sketch you could add a control "bouncing" as the push to increase the reliability of the readings. It is left as an exercise.


## MOTION SENSORS

This section will discuss specifically the "PIR" sensors (for "Pyroelectric Passive Infrared sensors"). Pyroelectricity is the ability of certain materials to generate a voltage when subjected to a temperature change. But beware, if your temperature (high or low) is kept constant, the voltage will gradually disappear.

What does this have to do with the movement? PIR sensors basically consist of two pyroelectric infrared sensors. And all objects emit infrared radiation, being further shown that the hotter an object, the more radiation such issues. Normally, both sensors detect the same amount of IR radiation (present in the atmosphere from the room or outside), but when a hot body as a human or an animal passes through the detection range, it will intercept the first one both sensors, causing a positive differential change from the other. When hot body leaves the sensitive area, the opposite happens: the second sensor which intersects the body and produce a negative differential change. These pulses are actually what the sensor detects. Thus, these sensors are almost always used to tell if a human has moved in or out of (usually large) range of the sensor: security alarms or automatic house lights are a couple of common uses for these devices.

The IR sensor is basically a FET transistor with a window sensitive to infrared radiation in its protective cover; changes in the level of IR light with a wavelength corresponding to body heat causes changes in the source-drain resistance, which is what the circuit monitors. Anyway, the big trick is an IR sensor incorporating lenses: their function is to condense a large area in a small, IR sensor providing a wider range of sweep. In fact, the quality of the lenses is basically PIR difference model of another sensor as the circuitry is common to all; otherwise changing a lens can change the amplitude and pattern of sensor sensitivity.


Besides all the already mentioned circuitry (IR sensors, lenses, etc.), a PIR sensor always has a chip which serves to read the output of IR sensors and process such that finally a digital pulse is emitted to the outside (that It is what our Arduino receive).

Most PIR sensors come in chips with 3-pin connection on the side or bottom: power, ground and signal data. The order of these three pins can vary by model, so you should check the datasheet (although most times each pin and is screen printed on the wafer own function). The food usually is 3-5 V DC but can be 12 V, so with a source of 5 V and V-9 work perfectly.

Other features common to most models features is that through its data pin emit a pulse HIGH (3.3 V) when motion is detected and emit a pulse LOW when not. The length of the pulses are determined by resistors and capacitors and present on the PCB are different from sensor to sensor. Its sensitivity range is usually up to a distance of 6 meters and an angle of 110 ° horizontally and 70 ° vertically. Most models integrate the chip BIS0001 to control the internal circuitry and sensor IR RE200B; the lenses can be varied. Concrete models of PIR sensors are for example Product No. 189 of Adafruit, No. 8630 SparkFun product or product 555-28027 Parallax, among others.

To test the PIR sensor, we can design a circuit like this. Keep in mind that the order of -food pin signal earth is different depending on the sensor type used. Specifically, shown in the diagram corresponds to the model PIR Adafruit, but the order of pins on the sensor

555-28027 is another (namely, if we look from the front, from left to right is the signal pins, power and ground) and also Sparkfun sensor (specifically, if we look from the front and from left to right, we have signal pins, land, food). In Sparkfun sensor it is also necessary to also connect a resistor "pull-up" of 10 k between the power pin and data pin to function properly; this implies that, unlike previous, Sparkfun sensor outputs a HIGH signal when no movement is detected and LOW when another signal.

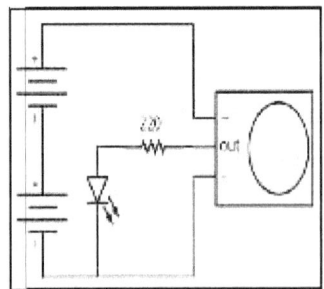

The idea of the circuit shown is that when the PIR sensor (assuming Adafruit model, which uses no external pull-up resistor) detects movement, the data pin will send a HIGH pulse of 3.3 V and thus illuminate the LED. When the LED is off, you can try your hand moving forward, directly or entire body.

Keep in mind that when the batteries are connected, it must wait between half a minute and one minute for the PIR to stabilize and begin to issue reliable data, so at first the LED may flicker slightly.

We must also make clear that the behavior of the PIR sensor Adafruit has the ability to operate in two modes depending on the position of a jumper located on the back of the plate. If placed in the "L" position, the LED will flash at a rate of every second while detects motion, and if placed in the "H" position, the LED will remain on while detects motion. He
sense of "L" mode is that (for instance) the PIR sensor can be connected to some other device that is activated when it detects a certain number of flashes of the LED.

Meanwhile, if we use the Parallax PIR sensor, you can specify how far we want to cover (in front of the sensor address) to detect motion. This is done by a jumper: if placed in the "S" position, movements that occur within a distance of 5 meters sensor is detected; if it is placed in "L" position, movements that occur within a distance of 9 meters (in this mode, however, false positives can occur) will be detected.

Example 7.26: Connecting a PIR (any of these) sensor to our Arduino is easy: only a pin-socket digital input pin must be connected to sensor data, for example # 2 (in addition to food 5 V and earth, obviously). The following sketch simply notifies the serial channel when motion is detected.

```
int disparo= LOW;  //No hay movimiento al principio int lectura = 0;
void setup() {
        pinMode(13, OUTPUT);
        pinMode(2,      INPUT);
        Serial.begin(9600);
}
void loop(){
  lectura = digitalRead(2);
  if    (lectura    ==    HIGH)    {
    digitalWrite(13,    HIGH);    if
    (disparo == LOW) {
      Serial.println("Movimiento detectado");
      //Solo queremos imprimir el mensaje una sola vez disparo =
      HIGH;
    }
  } else {
    digitalWrite(13, LOW);
    if    (disparo    ==    HIGH){    Serial.println("Movimiento
      finalizado");
      //Solo queremos imprimir el mensaje una sola vez
      disparo = LOW;
    }
  }
}
```

## The sensor ePIR

In Sparkfun (among other places) can acquire product code
9587 a somewhat different sensor called "ePIR" Manufacturer Zilog. The difference more important between this component and the PIR sensor is seen above that the former further includes a microcontroller within its own package. This allows greater flexibility to control the sensor and to manage the data.

Specifically, you can communicate with this component in two different ways: in "hardware mode" and "serial mode". In the "hardware mode", you can adjust the sensitivity of the sensor (ie, from which detected value is considered movement) or delay (ie, how long the sensor will wait after detecting movement to return to continue to detect new one), and other parameters.

The "standard mode", meanwhile, provides all the functionality of the "hardware mode" but also allows more advanced settings by sending specific commands (such as detecting motion in only one direction, to extend the detection range 3 x 3 m to 5 mx 5 m and more). For example, to force the detection of movement, we must send the command "a", which serves for the sensor reading as ASCII 'Y' character (for movement) or 'N' (or not). Anyway, for all the commands available and their possible uses, I recommend consulting sensor datasheet.

To select the "hardware mode" when starting the sensor (or exiting their low power mode) should provide a lower voltage of 1.8

V to the pin No. 4. Moreover, the specific value of this voltage determines the sensitivity of the sensor, which corresponds to 0 V and 1.8 V greater sensitivity to the child. Therefore, if the pin is connected directly to ground have the "hardware mode" and activated with maximum sensitivity. If what you want is to regulate such sensitivity hand, one option would be to connect this pin to the center pin of a potentiometer (potentiometer ends in this case should be grounded and power, respectively) using appropriate voltage dividers 0 to 1.8 V to obtain the desired range.

To select the "standard mode" when starting the sensor (or exiting their low power mode) should provide a higher voltage of 2.5 V to the pin No. 4. One way to achieve this is to connect resistance "pull-up" (typically 100k) between this pin No. 4 and the power supply (which attention must be 3.3 V).

The other pin of the sensor have a different value according to set work mode. In the "hardware mode", the required connections are:

Pin # 1: This pin is connected to ground.

Pin # 2: This pin must be connected to the power supply. this has to be between 2.7 and 3.6 V, so that the pin "3V3" on the Arduino board is ideal.

Pin # 3: This pin is used to specify the delay sensor (remember that's how long it will wait since the sensor has detected movement to be detected again). It is actually only an analog input which can receive from 0V (corresponding to a delay of 2 s) to 2 V (corresponding to a delay of 15 m), which could be connected for example to a potentiometer to regulate the tension applied by hand. If it is grounded, the delay will be fixed 2 s.

Pin # 4: This pin is used to select the type of mode (hardware or serial) you want to use. The specific procedure explained in the preceding paragraphs.

Pin # 5: This pin must be connected to a digital input of our Arduino. Through this pin one LOW value is received if the existence of movement, or a value HIGH if not detected.

Pin # 6: This pin is only an analog input you should get a

value proportional to the amount of ambient light tension. The idea is to activate the motion detection only in low-light environments (at night, for example). Specifically, when the voltage applied to this pin is less than 1 V, the motion detection is disabled. Therefore, normally, this pin is connected to a photoresistor, but can also be used a potentiometer to have a more direct control. If this functionality is not desired, simply connect this pin directly to the power supply 3.3 V.

Pin # 7: This pin is to be connected to a digital output of our Arduino. If this pin is LOW sends a signal, the sensor will go into a low power state ("sleep mode") and inactivated. This state is useful when the sensor is not in use. When you receive a HIGH signal is reactivated.

Pin # 8: This pin is connected to ground.

In the case of wanting to use the "standard mode", the required connections are:

Pin # 1: same use as the "hardware mode".

Pin # 2: same use as the "hardware mode".

Pin # 3: This pin must be connected to the TX pin of the Arduino (or one SoftwareSerial simulated with the library). Used to receive commands from this.

Pin # 4: also serve to select the type of mode (hardware or serial) you want to use (the specific procedure has been explained above), also serves to receive the readings taken by the sensor, so in addition it must be connected to the RX pin of our Arduino (or one simulated with SoftwareSerial library).

Pin # 5: This pin must be connected to the RESET pin of the Arduino board.

Pin # 6: same use as the "hardware mode". Pin # 7: same use as the "hardware mode". Pin # 8: same use as the "hardware mode".

Example 7.27: Here we show some sample code to use the "hardware mode", in which an LED (connected to digital output No. 12 our Arduino) lights whenever motion is detected:

```
int sleepModePin = 4;    //Donde se conecta el pin nº 7 del sensor int motionDetectPin =
2; //Donde se conecta el pin nº 5 del sensor int lectura;
void    setup()    {    pinMode(12,    OUTPUT);
    pinMode(sleepModePin,            OUTPUT);
    pinMode(motionDetectPin, INPUT);
    /*El pin " sleepModePin"  ha de recibir una señal HIGH
    para activar la detección de movimiento */
    digitalWrite(sleepModePin, HIGH);
}
```

```
void loop() {
    lectura = digitalRead(motionDetectPin);
    if(lectura == LOW) {   //Se detecta movimiento digitalWrite(12,
        HIGH);
    } else {            //No se detecta movimiento
        digitalWrite(12, LOW);
    }
    //Me espero dos segundos para volver a detectar movimiento delay(2000);
}
```

Example 7.28: Here we show some sample code to use the "standard mode", in which we have replaced the LED by a message read by the serial channel indicating whether motion has been detected or not:

```
#include <SoftwareSerial.h>
int sleepModePin = 4;   //Donde se conecta el pin nº 7 del sensor

int lectura;
/*El pin 4 es el RX de la placa (conectado al nº 4 del sensor) El pin 3 es el TX de la
placa (conectado al nº 3 del sensor)*/ SoftwareSerial ePir = SoftwareSerial(4,3);
void setup() {
        pinMode(sleepModePin, OUTPUT);
        digitalWrite(sleepModePin, HIGH);
        /*La comunicación con el ePIR en modo serie ha de ser
        siempre a 9600 bits/s.*/ ePir.begin(9600);
        Serial.begin(9600);
}
void loop() {
        //Ordeno    la    detección    de    movimiento
        ePir.print("a");
        //Mientras  no  se  reciba  respuesta  del  sensor,  no  hago  nada
        while(!ePir.available()) {;}
```

```
/*Muestro la respuesta del sensor:
' Y'  si  hay  movimiento y ' N'  si  no  */
Serial.print(ePir.read(););
//Me espero dos segundos para volver a detectar movimiento
delay(2000);
}
```

## Contact Sensor force

These sensors (also called FSRs, English, "Force Sensitive Resistor") to detect force. They are basically a resistor that changes its resistance depending on the force to which it is subjected (i.e., its resistance decreases more strongly received). These sensors are very

Cheap and easy to use but not too accurate: one size may vary between a sensor and another to 10%. So what one can expect from a FSR is to measure response ranges; ie although the FSRs are used to detect weight are poor choice for detecting the exact amount of this. Without

But for most touch-sensitive applications, such as "this has been squeezed certain amount" is an acceptable and economical solution.

The FSRs are composed of an "active" area (in a generally circular or square and in different sizes depending on the model), and two terminals, as this device resistance, are not polarized. Normally, they can withstand forces ranges from 0 to 100 newtons and the range of resistance offered range from infinite resistance when detected force to about 200 ohms at full strength. In the datasheet on the specific FSR we always find how that relationship is "applied force -> resistance obtained" which is not exactly linear (small forces because there is a large variation in resistance, and higher strength variation and it is lower).

90

To test it you can use a multimeter to resistance measurement mode. Pressing the sensitive area of FSR should be observed resistance changes.

7.29 Example: If we use this component in a circuit with Arduino, we connect a terminal to food and the other resistance "pull-down" (eg 10 k), which must be grounded. In addition, a point between the "pull-down" resistor (fixed) and the FSR resistance (variable) it must connect to an analog input of the Arduino board. It is the same assembly that we saw when we talk about LDRs and thermistors, actually. The idea is also the same as in the previous cases: read voltage that is in this point that increases as by simple Ohm's law, the resistance of the FSR decreases (ie, as is applied harder ).

Then sketch a very similar to other previously seen, which turns progressively as we go pushing LED plus a FSR is presented. In this example, the LED is connected (through its inseparable voltage divider) to the PWM output pin No. 11. The FSR is connected to the analog input pin number 0.

```
int lectura; int brillo;
void setup() {
        pinMode(11, OUTPUT);
}
void loop() {
        lectura = analogRead(0); //Cuanta más fuerza, más voltaje leído

        brillo = map(lectura, 0, 1023, 0, 255); analogWrite(11, brillo); //A más
        fuerza, más brillo delay(100);
}
```

Much like could perform another very interesting circuit consisting of a controlled servomotor through an FSR, so that the steering angle that was proportional to the force exerted on this. The trick would be to use map () to map the range of values read by analogRead 0-1023 () the range of values 0-179, which is supported by miservo.write () (function may have to replace the analogWrite () the previous code). It is left as an exercise.

If we want to know is actually the specific resistance value of the FSR, we can use in our sketches the following formula: Rfsr = (Rpull · 1023 / Vconvertido) - Rpull where Vconvertido is the value read by the analog input Arduino once transformed by the analog / digital converter, pull-up is the value of the "pull-down" resistor (fixed value) and Rfsr is the resistance value of FSR we want to know. This formula is identical to those obtained previously in the study of the photoresistors and thermistors as to obtain the reasoning is the same.

Knowing Rfsr could deduct the amount of force received by the sensor, but unfortunately there is no analytical formula relating the two variables, so we have no choice but to consult the graph datasheet for the exact relationship between the value of Rfsr calculated and the value of the corresponding force.

Example 7.30: The following sketch uses the formula mentioned in the previous paragraph for the resistance of the FSR. The circuit is the same as the previous sketch, but without LED: We will use the "Serial Monitor" to read different data.

```
int lectura;
const int rpull = 10000 //La resistencia pull-down es de 10KΩ
unsigned long fsr;   //Pueden ser valores muy altos unsigned long
cond; //Pueden ser valores muy altos long fuerza;
void setup() { Serial.begin(9600);

}
```

```
void loop() {
        lectura = analogRead(0);
        if (lectura == 0) {
                Serial.println("No hay fuerza");
        } else {
                fsr=(rpull*1023/lectura)              –              rpull;
                Serial.print(" Resistencia:              " );
                Serial.println(fsr);
```
/*La gráfica del datasheet muestra la fuerza en función no de la resistencia, sino de su inversa, la conductancia, medida en microMhos.Por eso se ha de invertir convenientemente el valor de
fsr*/
```
                cond = 1000000 / fsr;
```
/*Y ahora usamos los valores de la gráfica para aproximar el valor estimado de la fuerza. Dependiendo del modelo concreto de FSR los valores en las condiciones de los " ifs" deberán ser diferentes */
```
                if (cond <= 1000) { fuerza = cond / 80;
                        Serial.print("Fuerza      en      Néwtones:      ");
                        Serial.println(fuerza);
                } else {
                        fuerza = (cond – 1000) / 30; Serial.print("Fuerza
                        en Néwtones: ");
                        Serial.println(fuerza);
                }
        }
        delay(100);
}
```

7.31 Example: A curious example of application of FSRs (or indeed any other analog sensor) is the next circuit. It FSRs have 3 connected in parallel to the analog input pins on the Arduino board # 1, 2 and 3 and to ground through a "pull-down" resistor 10 k. We also have a buzzer or speaker connected to digital pin exit 8 (through a voltage divider 100 ohms).

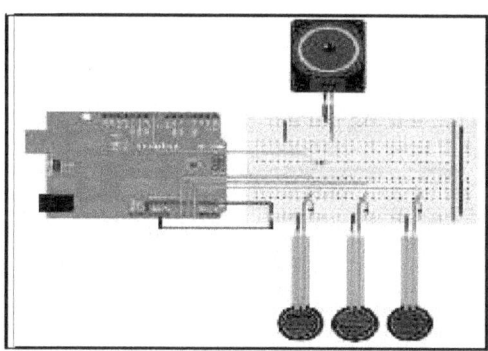

The code does is read the readings of three sensors, each of which corresponds to a musical note in an array of notes. If any of these sensors is pressed beyond a threshold, the corresponding note sounds. We have therefore implemented a simple electronic piano.

```
//Más allá del umbral sonará la nota const int
umbralminimo = 10;
   //Cada nota corresponde a un sensor int
   notas[] = { 220, 440, 880 }; void setup(){}
   void loop() {
      int i;
      int lectura;
      //Voy recorriendo los sensores uno tras otro for (i = 0; i < 3;
      i++) {
         lectura = analogRead(i);
```

94

```
            //Si se presiona lo suficiente if (lectura >
        umbralminimo) {
                tone(8, notas[i], 20);
            }
        }
    }
```

In Adafruit FSR distributed model product number 166 and sold Sparkfun several different features: No. 9375 and No. 9376, but they are all manufactured by Interlink (http://www.interlinkelectronics.com).

## Flexion sensors

Some like the FSR sensors are flex sensors (in English called "flex sensors" or "sensors bend"). These sensors are made of a flexible resistive strip in one direction. Its resistance changes depending on how much is arched: If they are balanced (ie, no warping) resistance is minimal and

It offers more resistance the more flexed.

As the FSR, have two terminals, one can connect within our circuits to the power supply (preferably via a voltage divider) and the other to a "pull-down" resistor which is grounded (also you could use alternative configuration using "pull-up" s). At the point where the sensor to the "pull-down" resistor is connected you must connect an analog input of the Arduino to read the resulting voltage at that point. As we know, this value depends on the value of the resistance of the sensor, so it will help us to know how much is flexed.

7.32 Example: The following sketch shows a very basic example of usage:

```
void setup () {Serial.begin (9600);
}
```

```
void loop () {
int sensor, degrees;
sensor = analogRead (0);
/ * Read I convert the value to degrees of flexion. The first two numbers of map () (768 and 853)
are the values read when the sensor is completely straight and when you have a curvature of
90 degrees, respectively. These values can be accessed in the datasheet or if any previously
calibrated. The next two numbers map () are the degrees to which we want to map (0 degrees
and right angle) / *
degrees = map (sensor, 768, 853, 0, 90);


Serial.print ("The degrees of flexion are:"); Serial.println (degrees, DEC);
delay (100);
}
```

Sparkfun distributes several models of different lengths: the product No. 10264 and 8606 are two examples. Adafruit only distributes the second with code 182. In addition to length, other important features to consider are its width and weight, since often these sensors are used in textile projects.

Do not confuse the flexion sensors with sensors called "FlexiForce" which is a trademark of Tekscan (http://www.tekscan.com) brand. The "FlexiForce" FSR sensors are sensors and Sparkfun provides few supporting different power ranges (product No. 11207, 8685, 8712 or 8713, etc.).

Shock sensors

Because of their internal electrical constitution, buzzers also can furthermore be used as sound emitters as shock sensors. The mechanism is just the reverse of conventional: shock (soft) received by the buzzer result in internal vibration plate, which generates a series of electrical pulses that can be read by an Arduino board. Thus, we can design circuits that respond to touch and even distinguish the exerted pressure. As the buzzer is an analog device, according to how strong is hit, the read signal from the Arduino board will be greater or lesser intensity.

Example 7.33: Knowing this, we can design the circuit example shown below, where we have a speaker and two buzzers. As you can see, the speaker is grounded and no digital output 8 (a through a series resistance of 100 ohms), the positive terminal of the buzzer (if these are polarized) are connected to the analog inputs No. 3 and # 5, and the negative terminals of the buzzer (if these are polarized) to ground. Furthermore, each buzzer is connected in parallel to a resistor (a recommended value 1 M), whose function is to act as resistance "pull-down", maintaining the analog input to 0 V while the buzzer is not pressed.

If we have no buzzer hand, we can achieve the same effect acquiring a piezoelectric sheet such as Product No. 10293 Sparkfun, which is nothing more than a buzzer without coating or acquiring the Product No.

10772, consisting of four sheets kit as above plus several resistors of 1 M. We can also obtain the "Sound & Buzzer Module 'from Freetronics, which is merely a convenient breakout joined plate buzzer.

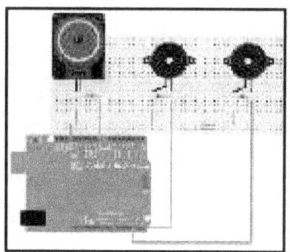

From this circuit, we can write code like the following. The idea is that as you press the buzzer will sound from the speaker a note or another. In addition, the stronger we tighten a buzzer, your note will sound longer. This is possible because we know the pressure on the buzzer: they reach an amplitude of vibration depends on the magnitude of the blow: the more pressure, more amplitude. And the results proportionally amplitude voltage, voltage detected by the analog inputs on the Arduino board. Just as the buzzer is an analog device, the amplitude of vibration after suffering a stroke (ie, the voltage level received) will increase gradually, will peak and decrease again to stop. Therefore, the trick to detect the magnitude of the coup is not really detect the maximum voltage obtained (which may be a not very accurate data) but check how long the voltage received above a certain threshold is maintained (decided by us): the longer the time between the first and last time a voltage value greater than the threshold, the greater is detected will have been the magnitude of the blow.

```
int umbral = 100; //El zumbador se considera pulsado sobre ese umbral int lectura1
= 0; //Lectura obtenida de un zumbador


int lectura2 = 0; //Lectura obtenida del otro zumbador
int tiempo = 0; //Tiempo en que las lecturas son superiores al umbral int do = 1915;
//Semiperíodo de la onda de la nota " do"
int re = 1700;   //Semiperíodo de la onda de la nota " re"
void setup(){
        pinMode(8, OUTPUT);
}
void loop(){
        //Se  detecta  si  está  presionado  un  zumbador
        lectura1= analogRead(3);
         if (lectura1 > umbral) {
                tiempo=0;
/*Mientras se detecte que la presión continúa, se sigue leyendo la entrada analógica
para ver si aún se lee un valor por encima del umbral. El tiempo transcurrido hasta
que se lea un valor menor al umbral marcará la duración de la nota a escuchar*/
                while (lectura1 > umbral) { tiempo++; }
/*Para evitar posibles ruidos no deseados y considerar el golpe válido, se establece
un tiempo mínimo de presión */
                if (tiempo > 100) { sonido(do, tiempo); }
        }
```

```
        //Se detecta si está presionado el otro zumbador lectura2=
        analogRead(5);
         if (lectura2 > umbral) {
                tiempo=0;
                while (lectura2 > umbral) { tiempo++; }
                if (tiempo > 100) { sonido(re, tiempo); }
         }
}
void sonido(int nota, int tiempo ) {
         unsigned long duracion;
         duracion = micros() + (35000 * tiempo);
/*Empiezo a contar desde el momento actual y añado un valor arbitrario
multiplicado por el valor de " tiempo" ; así, cuanto más fuerte se pulse el zumbador,
más durará */
         //Mientras    no   se   llegue   al   final
         while(micros() < duracion){
             digitalWrite(8,              255);
             delayMicroseconds(nota);
             digitalWrite(8, 0);
             delayMicroseconds(nota);
         }
}
```

## Sound sensors

Actually, a sound sensor is merely a pressure sensor that converts air pressure waves (sound waves) into electrical signals of analogue type; It ie a microphone. Many types of microphones as the physical mechanism used to perform this conversion: those of "inductive" (also called "dynamic"), the "capacitor", piezoelectric, type etc. Depending on the type, some have a better response to a range

determined sound frequencies other (ie, to be more "faithful" to the

original waveform), some will have higher sensitivity than other (ie, already generate a given voltage to minor variations detected volume), a start distorting at lower volumes than others (ie, which provide a lower THD for a given voltage), some will be stronger and more durable than others, etc.

However, in our projects with Arduino UNO plates variety of microphones to choose it is drastically reduced. Arduino UNO is not designed for audio processing platform: we have seen that (although there are outstanding projects in the field of synthesis) the generation and emission of sound is certainly limited. And the same goes for the sound reception: first, the female pin-plate are not capable of receiving AC (which is what the audio signals). In addition, the analog-digital converter takes at least

100 microseconds to perform a reading input, so that the maximum possible sampling frequency is 10 KHz (ie, a relatively low quality). In addition, the processing of an acoustic signal (actually composed of a set of multiple analog signals of different frequencies and amplitudes) is much more complex than the ATmega328P limited memory and is able to perform with reliability.

Therefore, the microphones used together with Arduino ONE plates in most cases are only used as simple detectors and / or sound volume (or at most, connecting them to special chips as MSGEQ7 Mixed Signal Integration They could be used as detectors of sound frequencies, but this possibility will not be discussed).

Specifically, we will use a variant type condenser microphone called "electret microphone." This type of microphone is omnidirectional (ie, detects the sound from all directions and not only from a particular point) and usually have good sensitivity (ie, the generated voltage varies greatly accordingly as to vary the intensity of sound) . In addition, they are cheap and small size. Have better frequency response medium-high range, so they are best for voice communications that important rhythm section of music, for example. In many household devices such as mobile phones, computers or headset microphones they are built are of this type.

Whatever the practical application to give you a microphone, in any case its use necessarily involves the use of a pre-amplifier. This is because the signal generated by a microphone has too small amplitude (typically between 0 and 100 millivolts) to be used by our Arduino. Therefore, we can not connect a microphone directly to an analog input of our Arduino as we have been doing with other analog sensors, but we will include a pre-amplifier between the microphone and the Arduino. In this sense, we must distinguish between pre-amplification (conversion of the signal generated by the microphone to take her to a level usable by the circuit (in this case, Arduino) and amplification (conversion of the usable signal internally in our circuit an audible level enough to broadcast it through speakers).

In fact, this process of pre-amplification and amplification is also necessary when you connect the microphone to a system of home and professional audio: all intermediary devices (from the portable file "mp3" or similar to hi-fi going by televisions, DVD players, mixers, etc.) work at levels higher than the voltage provided by any microphone (the so-called "line level" turn different from the one used by Arduino) so that the signal generated this must be pre-amplified to be used by all these audio devices. When this electrical signal is already at line level, to transform it into actual sound and be able to broadcast on a speaker system is necessary then amplifying it to the desired power.

## Breakout plates

There breakout plates incorporating an electret microphone and preamplifier in one, so that we can start using the full kit microphone pre-amplifier + instantly. One example is the "Sound Microphone Input Module 'from Freetronics. The insert has four connectors: "VCC" (to connect to

the supply of 5 V provided by our Arduino), "GND" (to Earth), "MIC" (analog output to connect to an analog input of our Arduino) and "SPL" (another analog output to connect to another analog input of our Arduino). In our projects we can use the signals received by both channels ("MIC" and "SPL") but usually connect and use only one of them, as we are interested: the channel "MIC" provides the most faithful amplified signal acoustic signal possible, so it is most useful when you want to perform audio processing; however, the "SPL" channel ("Sound Pressure Level") simply provides a voltage which is proportional to the volume of sound received, so it will serve to easily detect the existence of sound and "quantity", which is generally we will want.

7.34 Example: The following sketch reads the SPL value from this plate five times per second and displays the reading by the serial channel: more read voltage, the sound volume.

```
//La salida SPL está conectada al pin de entrada analógico 0 const byte
splSensor = 0;
  void setup() {
    Serial.begin(9600);
  }
  void                loop()                {
    Serial.println(analogRead(splSensor));
    delay(200);  //Evito la sobrecarga del canal serie
  }
```

A plate similar to the above but only provides an output type "SPL" is the number of DFRobot DFR0034 product. Another similar plate which also provides only SPL output type is Inex ZX-Sound of Robotics.

On the other hand, there are plates whose output is digital: if they detect a sound above a threshold (generally defined via a potentiometer), send a HIGH signal to the digital input of the Arduino where they are connected, and if the sound detected is below that threshold, send a low signal. These inserts are useful when you do not want to monitor the volume of the environment, but only certain sounds with a higher volume remaining (a practical application is to detect collisions in robots, for example). An example of this style is insert the Parallax Product No. 29132; only it has three connectors (5

V, GND and signal) and by a potentiometer that incorporates can calibrate the sensor sensitivity, so that the threshold separating sound sending a HIGH signal (noise is detected) of a LOW (noise is detected) is

conform to our needs. Another similar plate is called "Sound

TTL sensor output "of Cutedigi.

Example 7.35: Here are some sample code very easy to use for both chips:

```
int lectura = 0;
void setup() { Serial.begin(9600); }
void loop() {
   //La plaquita está conectada a la entrada digital nº 2 de Arduino lectura =
   digitalRead(2);
   if (lectura == HIGH){ //Si se detecta un sonido más allá del umbral
      Serial.println("Sonido detectado");
      delay(100);
}
```

A plate which provides an output only type "MIC" (ie, an electrical signal that reproduces the behavior of the acoustic wave) is the product no

SparkFun 9964. Once connected to this output "MIC" to the analog input of the Arduino (plus connector "VCC" a source of 5 V and "GND" connector to common ground), if we look at the values received by the "Serial monitor "see that when no sound is detected remained stable around the value 512 and in noise fluctuate above and below the central value (both being higher this deviation above as below-the larger the volume of the noise detected). Therefore, if we get the volume (ie, use this as a simple sensor plate SPL), we should use the expression volume = abs (analogRead (0) -512); (Assuming that the analog input used is pin-female No. 0). But why this behavior?

Because what makes this insert is "move" the value of the corresponding output to 0 V (the central value of the wave) to a central value of 2.5 V and with it, the rest of blocks of securities without altering therefore the waveform. In other words: add a "buffer" of 2.5 V (DC) on which the audio signal travels (AC). This is what is called having an "offset" (or "biased" in English) signal. The aim is to make the Arduino can detect both positive values and negative acoustic wave. Without the offset signal, negative values actually be below 0 V and then the analog input of the Arduino (which only supports DC power, not AC) could not detect correctly. With the offset signal, the positive peak may reach a maximum of 5 V (ie, may have a maximum amplitude of 2.5 V from the 2.5 V base offset signal) and the negative peak to a minimum 0 V (ie, you can have a maximum range of -2.5 V signal from off-base

2.5 V).

Another trick you can use with an output type "MIC" (both wafer Freetronics like SparkFun) is to distinguish whether the detected sound is a sound point (a bang, for example) or a continuous sound (like a conversation, for example). Because the Arduino UNO is relatively slow to take samples, it may not detect all changes in real-time volume and "leave detect" peaks of sound that appear in between two readings. To address this issue, it can be coupled to the output "MIC" circuit generically called "envelope detector" (in English, "envelope detector"), shown below.

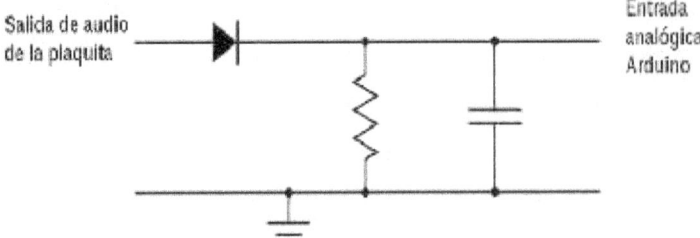

The purpose of this circuit is to get to the Arduino only the extreme values of each positive peak, linking them together in a gentle way, regardless therefore all the vibration of peak to peak. The figure of the received signal (the "envelope") help the Arduino to "go along" changes in volume: if the envelope contains narrow peaks will have detected a sharp noise, and if the contrary, we are detecting a sound more or less continuous. But why this circuit works well?

On the positive peaks, the signal passes the diode, charging the capacitor to (almost) the peak value. Negative peaks in the diode is reverse biased, allowing no current to flow, so that the capacitor will react slowly discharged through the resistor (Arduino input is designed to attract a number of negligible current). When he reappears a positive peak, the diode again miss signal and the capacitor reloads quickly, and so all the time. The rate of discharge of the capacitor is defined by the product R · C (called "time constant"), and its value depends on the type of project we want to perform, but a typical value is 0.1 start testing.

Another thing we can do with the output "MIC" is turning into an outlet "SPL". In the case of Freetronics plate we will not be necessary because we already have the two types of output, but in the case of the insert Sparkfun, the trick is coupled to the output "MIC" circuit generically called "fixer positive level "(in English," positive clamper unbiased "), shown below.

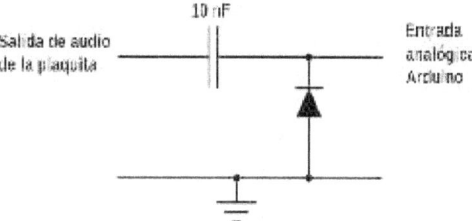

The objective of the above circuit is decentralize the AC signal such that, without changing the shape of the wave, the ends of the negative peaks are always just a minimum value of 0 V. If the volume of sound changes, the length of the also signal peaks, but as what is fixed is that the minimum value is always

0 V, what will happen is that the central value should be moved to meet this condition, which will change the end value on the other hand, the positive peak: when the volume increases, the extreme value of the positive peak will increase as the volume decreases, the extreme value of the positive peak also decrease. And this is what we measure. It is recommended that the diode is type "germanium" (the 1N34A would be a good example) and in this case, is also recommended to feed the plate with 3.3 V instead of 5 V. If this is done, the values received by the analog input of the Arduino range between 0 and 750/800 (more volume, greater maximum value received). We can then write a skit to react to sounds above a certain threshold, for example, simply observing the maximum analog values received.

**Pre-amplifier circuits**

We may dispongamos a single electret microphone instead of a whole wafer breakout. If we connect to our circuit, the first thing to know is that the electret microphones (such as Product No. 8635 Sparkfun, for example) should be fed. They are also polarized devices, in which the negative terminal is often marked by a notch or signal. In principle this should be negative terminal grounded and the positive terminal should be connected to a power supply, which can be any capable of providing a voltage between 2 V and 5 V (if necessary, through

475

a voltage divider). The audio signal received from the positive terminal would get, always through a condenser (from 0,1μ F and 1μ F) acting as a highpass filter to remove any DC mattress AC signal obtained.

Unfortunately, connect an electret microphone is not so simple because we know that it is necessary to pre-amplify the received signal to make this usable. Therefore, we must build an accessory circuit around him to perform this function. A simple example is the following, in which we have only used a few resistors and an NPN transistor (such as 2N3904), whose function is precisely amplify the signal, as studied in the previous chapter under the heading of sound generation by tone ()):

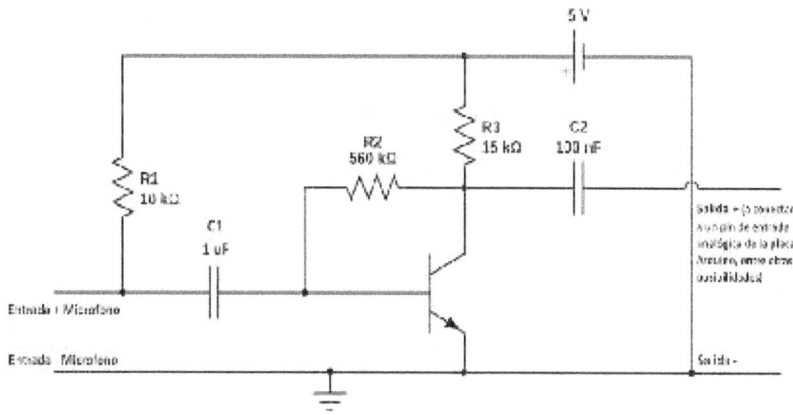

The resistor R1 serves as voltage divider electret microphone and capacitor C1 serves, as already mentioned, to eliminate possible DC AC mattress received signal. From here, the operation should know: the signal received by the microphone is sent to the base of the transistor, which will leave more or less current flow between collector and emitter depending on the intensity of that. If we take the current generated as the output of the circuit, we have a modeled representation of the original signal but wider.

The role of resistors R2 and R3 is decentralize the AC signal received by the base of the transistor from the microphone. In other words: added to that signal a "cushion" constant DC that allows all the values of the wave AC

60

(Including negative peaks) above 0 V. In the absence of sound, the base of transistor receive a certain current DC transistor remain in a state of permanent intermediate driving called "point-Q"; when sound is detected, the oscillations of the signal will fluctuate around that intermediate state of driving without ever reaching saturation mode or to one side or the other cutting mode (because the signal is self-regulating: if the current through the collector increases, decreases the current in the base, and thus automatically the current passes through the collector reduced). Furthermore, due to the emergence of this new mattress DC, after obtaining the necessary eliminate extended signal as far as possible through the capacitor C2.

The values of R2 and R3 should be chosen with care, because they mainly depend amplitude of the amplified signal, which must be one or the other as we interested (line level voltage of Arduino, etc.). In fact, the most common use of this circuit (and all preamplifiers) is to adapt the signals to line level so that they can be derived amplifiers audio circuits themselves.

If we connect this preamplifier our Arduino circuit and read the values of R2 and R3 shown in the diagram above, we can see through the "Serial monitor" the presence of a DC mattress 320 (about 1024) and peaks up 950 on very loud sounds. These values provide a useful quite acceptable range (within the range 0-1023 supported by the analog inputs on the Arduino board), but if you want, you can try other values of R2, R3 and C2 to improve (ie, to reduce plus the value of the mattress and further increase the useful range within accepted).

7.36 Example: To test the above pre-amplifier circuit can run the following code, which can serve, in a quiet environment to observe fluctuations in the signal present even though in theory we should have a constant signal. To avoid this phenomenon, we can use the application to calculate the average of the last read values (as we have seen in previous examples) and thus soften the unwanted peaks.

61

```arduino
void                  setup(){
Serial.begin(9600);
  }
  void loop(){
    int mini = 1024; //Iré reduciendo el valor de " mini"  hasta el real int maxi = 0;
    //Iré reduciendo el valor de " maxi"  hasta el real




    for(int  i=0;i<1000;i++)  {  //Otra  posibilidad:  while(millis()  <  1000) int valor  =
        analogRead(0); //El micro está conectado en pin 0 mini = min(mini, valor);
        maxi = max(maxi, valor);
  }                   Serial.print("Ruido=");
    Serial.println(maximum-minimum);
  }
```

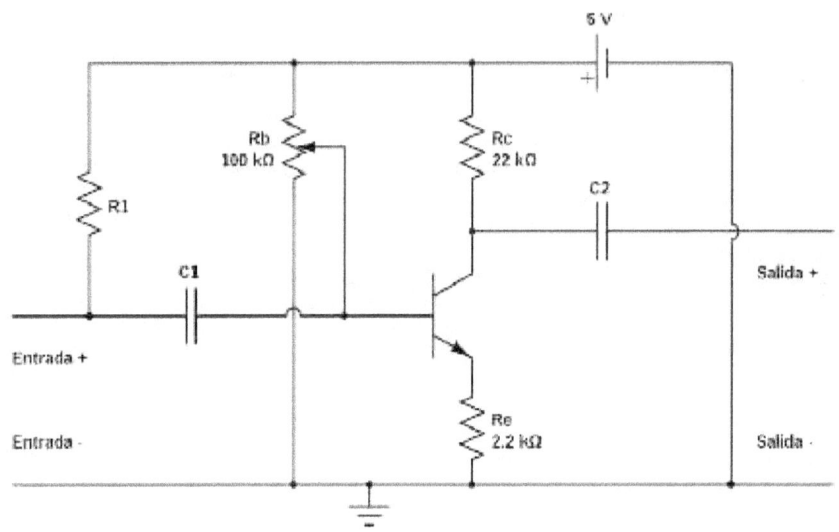

63

Therefore this circuit as the above have the advantage of amplifying the signal without distortion, but has the disadvantage that constantly provides DC mattress to the base of transistor, which is not energy efficient circuit.

Finally, I would not stop commenting that in many Internet forums and blogs are suggested as a pre-amplifier chip quite popular, the LM386. It is not a good choice, because in reality, the LM386 amplifier chip is working to line level. That is, it is designed to provide, from an entry

Audio and pre-amplified power up to 1 W (in its most capable versions) to a speaker system. If you connect directly to a microphone, we get too much noise; a better alternative would be to use in any case the LM358 chip.

Nor is it a good idea (as is often proposed) use the LM386 as amplifier by connecting directly to an audio output of our Arduino, because, as stated above, this chip is designed to have input line level, which allowed lower voltages than the 5 V offering a pin-female Arduino.

Speech recognition

Our Arduino is able to respond to commands by voice expressed thanks to "EasyVR Shield" from Veear (distributed among others by SparkFun with product No. 10963), which includes a voice recognition module designed and manufactured by the same company Veear ( autonomously also available). This shield also includes a microphone, an output for connecting an 8 ohm speaker and 3.5 mm jack socket for connecting headphones.

This shield is connected to the Arduino via serial channel (preferably by two RX and TX pins defined by software) at a rate of 9600 bits / s. Although we could control this shield sending the appropriate commands from a sketch, to manage its functionality and performance it is much easier to use the library which the manufacturer offers http://www.veear.eu/downloads. The necessary documentation to learn its use is included in the downloaded package.

This factory shield includes a set of predefined commands available in several languages (including Spanish), ideal for basic controls, but also supports the definition of own fully customized up to 32 commands, including passwords. To define these commands and / or set prerecorded commands you need to use the graphic software (called "EasyVR Commander") provides that the manufacturer (though only for Windows system) http://www.veear.eu/downloads. The information needed to learn to use the can consult the User Guide, available in the same download site.

## COMMUNICATION NETWORK

In this chapter we will see different ways to communicate our Arduino with other boards (or computer) connected to networks of different types: specifically Ethernet wired networks, Wi-Fi networks and Bluetooth networks. The goal is to control and transfer information between these devices remotely. So, we could, for example, access data from sensors or actuators control facility without having to physically move. To do this we assume, unless the contrary, we use either the Arduino Ethernet board or the Arduino UNO with Arduino Ethernet shield coupled do not tell.

Networking basics IP Address
One thing that always has to be assigned a plate / Arduino Ethernet shield
so that it has connectivity to the network it is an IP Address. In fact, any device (such as a computer) have been properly configured its own IP address to be part of a TCP / IP network.

The IP address is a numerical label consists of four figures, values between 0 and 255 separated by periods, that identifies the network card of a device (computer, Arduino Ethernet board, etc.) within the network TCP type / IP. Each card has a unique IP address, so using these addresses
the devices can recognize and communicate with each other. An example could be ip 192.168.0.1.

In a computer, the IP address can be manually set by the user (which is called using a "fixed IP" or "static IP") under different specific, different utilities depending on the operating system used, or may exist on the network automatically grant a specialized ip addresses to other devices when they request it (what is called using a "dynamic IP") device. This request can be made by decision of the user or, more often, automatically when you boot your computer using a protocol called DHCP message exchange. When this protocol request-granting ips used, the device gives the ip is often called "DHCP server" and the device that the request is called "DHCP client". Both types of IP (fixed or dynamic), regardless of how they are established, are functionally identical.

The Arduino Ethernet (and similar) plate can purchase your ip also of two ways: either permanently establishing its own specific value within the code of our sketch, either dynamically either from an existing DHCP server on the network. In any case, we have to use the official "Ethernet" library.

It is beyond the scope of this book detailing how to configure the ip permanently on computers running different operating systems. Refer to the official support of each system. Likewise, neither will detail how you can install and manage a DHCP server. Just for reference, discuss the existence of some applications for converting a computer in a DHCP server: On Linux you can use the "ISC Dhcpd" (http://www.isc.org/software/dhcp) the "Dnsmasq" software or (http://www.thekelleys.org.uk); Windows can use the built-in factory comes in versions Server or the DhcpServer (http://www.dhcpserver.de/dhcpsrv.htm) or "Open Dhcp Server" (http://sourceforge.net/projects / dhcpserver) or also the "Dual dhcp- Dns Server" (http://dhcp-dns-server.sourceforge.net), among others. Refer to its respective official documentation for how to set them up.

Netmask

The network mask used to identify which network a device that has a specific IP address belongs. A device (for example, our Arduino) can only belong to a certain time to a single network. Know which network a device belongs is very important, because only devices on the same network are able to communicate with each other.

A network mask is a set of four digits of values between 0 and 255 separated by a dot. There are many types of network mask, but we will focus on the three most basic: mask class (whose value is 255.0.0.0), the Class B (whose value is 255.255.0.0) and Class C ( whose value is 255.255.255.0).

To find out which particular network device belongs to a particular ip and mask, we know its network identifier, which is also a set of four numbers between 0 and 255 separated by a point. This identifier is formed by choosing the part of the IP device that matches part of your mask value 255, and then adding 0 to the party coincides with part of his mask to 0. For example, if we have a plate Arduino (or a computer, is the same) that has the ip 192.168.23.1 mask 255.255.0.0 and the network to which it belongs will be "192.168.0.0" and can communicate with all devices belonging to that same network (for example, always assuming the same mask, which has an IP like 192.168.142.62 or 192.168.216.39, etc.). Instead, the device can not communicate with another such as have the ip
192.76.23.123 (and the same mask), because this would belong to the network "192.76.0.0", which is different.

In a computer, network mask can be set manually by the user using the same applications needed to establish the value of a "fixed IP" (which we have said that depending on the operating system used are different), or may be assigned by the existing DHCP server on the LAN. On a plate / Arduino Ethernet shield, the mask can be set by writing "by hand" within the code of our sketch (using the "Ethernet" library), or be assigned through an existing DHCP server on the network.

Private IP addresses

When assigning an IP to our plate, we can come the question of whether any combination of four numbers is valid. The answer is no. For starters, each of the four numbers can only have a value between 0 and
255, but are even valid all possible combinations of these values:
we can only use an ip it "private" type. Let us explain this.

Each device connected directly to the Internet has a different "public" ip that lets you communicate with other devices in the world. The IANA (http://www.iana.org) is the international body that assigns these public ips controlled entities that need to have direct Internet access (such as telephone operators, etc.). But only works with the IANA internationally recognized organizations: public ips never granted to end users. So we as users can never manage public ips. In fact, the domestic Internet access is possible because we have hired the operator offers a public ips he has obtained from the IANA.

However, it is usually quite common in a company or organization (or even in a private home) all have multiple computers and want to connect with each other and the Internet. Plus it would be a great waste assigned to each of these teams a public ip (more considering that these are not infinite and that are currently running), we just say that this is impossible because the IANA does not allow this. The solution is to use "private" ips. That is, ips can function only within a local network, unable to "go out" (ie without access to the Internet directly).

In fact, the Internet itself is possible, although all the teams in our LAN using a private ip, thanks to the existence of an intermediate computer (commonly called "router") is the only one that is assigned an IP public and is used by other computers on the LAN as a gateway to the outside. With this "trick" we got multiple machines that have access to the Internet using a single public IP, camouflaging all inside the LAN.

The beauty of this system is that, in addition to saving public ips, private ips can be reused as often as desired in different LANs, because the latter are not seen each other: they are confined within the LAN. This allows us to assign teams to our organization a private ips and another person in any other part of the world assign the same in their own organization, no problem: there will be no conflict because the two organizations only their IPs are respective public (granted by their respective operators) and therefore there is no overlap.

So, in short: the only ips we can assign to our Arduino to communicate with machines on our local network are private ips. Several networks officially reserved exclusively for private use. The particular network you choose does not matter, but in any case all our LAN devices must belong to the same network so that they can "see" each other. Possible private networks to choose from are:

The network "10.0.0.0" class A (255.0.0.0): That is, in our devices can use ips spanning from the 10.0.0.1 to 10.255.255.254 (the   It is a special ip 10.255.255.255 and our projects will not use). All these computers belong to the same network (the 10), and can easily check counting possible IPS that it may be thousands of computers.

Networks "172.16.0.0", "172.17.0.0", "172.18.0.0" ... until the "172.31.0.0" Class B (255.255.0.0): In each of these networks can use to go from ips xx0.1 to xx255.254 (ie, from 172.16.0.1 up
172.16.255.254, or from 172.17.0.1 to 172.17.255.254, etc). It can be seen
we can easily use up to 16 different private networks Class B (as opposed to the only private class A possible), but in each of these networks of class B may be fewer teams.
Networks "192.168.0.0", "192.168.1.0", "192.168.2.0" ... until the "192.168.255.0" Class C (255.255.255.0). In each of these networks we can use ips spanning from xxx1 to xxx254 (ie, from
192.168.0.1 to 192.168.0.254, or from 192.168.1.1 to 192.168.1.254,
etc). You can see that we can use up to 256 different private class C networks, but each can only exist until
254 teams.

Mac address

Another essential for the Arduino Ethernet board can connect to the network, and an IP address and network mask, a fact that I have a MAC address. In fact, any computer must always have its own MAC address specified to be part of a TCP / IP network.

The MAC address is a label of 48 bits (12 hexadecimal characters) that identifies the network card unique and unambiguous way in the world. This figure does not depend on connection protocol used or the network; it is a value set by the manufacturer of the card (usually) can not be changed because I is recorded on this hardware. Fortunately, most devices (as in the case of computers) is not necessary to know (and even less change) the MAC address or to set up a home network or to configure the Internet connection or anything, because this one It used to more internal levels of the network and is preset at the factory. An example of MAC address could be 12-AB-56-78-90-FE.

In the case of the / Arduino Ethernet shield plate however, we do need to specify the code of our sketch its MAC address (by Seller
"Ethernet"). Depending on the age of the model plate / shield you have, you might not have the default or factory MAC address. If it is so, this will show up MAC printed on a plate attached to / shield label and the code of our program we use that MAC address. If we see no label, plate / shield will have no predefined MAC, so the code of our sketch we invent us (trying not match any other devices that are connected to our local network at that time) .

DNS servers

A DNS server is a computer (usually publicly accessible through the Internet) that enables users to use descriptive names instead of IP (harder to learn and remember) to identify and connect with other computers on the network addresses . That is, computers are located in different parts of the world that allow users to use a single name (such www.rclibros.es) to connect to a specific computer instead of typing its IP address (as 82.98. 148,182).

When a user enters a DNS name in an application of our computer (like a browser), the first thing that happens is that application consultation which DNS server (or servers, as there may be several) is preset to the operating system used. When he discovers that saved the ip default DNS server, it sends a query requesting to know what the actual IP address that corresponds to the name typed by the user. If the DNS server responds with the requested information, the application can then communicate directly with the remote computer using its IP address. If the DNS server has no entry in its database for the queried name, usually he consult another DNS server until you find one that does know the correspondence name <-> ip, or until it is found that the queried name does not belong to any existing machine.

In a computer, the DNS servers to query by default can be set manually by the user using the same applications needed to establish the value of a fixed IP or mask (which we mentioned that according to the operating system used are different) or they can also be assigned by the DHCP server on the LAN exists. In the case of plate / Arduino Ethernet shield, the DNS server we want to use can be set by typing its ip "manually" within the code of our sketch (using the "Ethernet" library) or can be assigned through an existing DHCP server on the network.

We must clarify that if a device (plate / Arduino Ethernet shield or computer) does not have a DNS server configured, you can still communicate with other computers by using their IP addresses directly. That is, the use of DNS servers is not technically necessary, but no doubt greatly facilitates the use of the network by users.

72

Some of the public DNS servers that can use in our Arduino sketches may be those of Google (with ips 8.8.8.8 and 8.8.4.4), the OpenDNS (208.67.222.222 and 208.67.220.220), those of DNSAdvantage (156 154. 70.1 and

156.154.71.1) or of ScrubIT (67.138.54.100 and 207.225.209.66), among others (As provided by each network operator).

You can install in our own LAN specific computer software to convert a DNS server, such as the "ISC Bind" (http://www.isc.org/software/bind) the "Dnsmasq" (http or application: // www.thekelleys.org.uk), but this will have to do unless we need that our local network computers also have names. Typically, DNS servers typically use the Internet are public (as listed in the previous paragraph), and to use the names of all public Internet computers (web servers, mail servers, etc.), which is which generally we will want.

Default Gateway

A default gateway (also called "gateway") is a specialized communicate two or more networks together (that is, connect and redirect data traffic between them) device. Generally, in homes or offices, this device (which is commonly called a "router" or router) connects the local network with the Internet address. In business, often this function falls on a computer that also redirect the data traffic between the local network and the external network (Internet), performs more tasks (like making firewall, for example). In any case, a standard "router" must internally add a network card with a public IP address assigned by the telephone operator (which will serve to identify within the Internet), and another card with a private IP network (which serve to identify within local network so as to be accessible to other computers on that network).

All computers on a local network must be configured private ip of default gateway for them to know where to direct messages destined abroad. In the case of computers running different operating systems is beyond the scope of this book detailing how this is done: I refer you to the official support of each system. In the case of plate / Arduino Ethernet shield, the default gateway that want to use can be set by typing its ip "manually" within the code of our sketch (using the "Ethernet" library) or can be assigned to through an existing DHCP server on the network.

We must clarify that if a team has configured a gateway, you can still communicate with other computers in your own local network, but when you need to send a message to the outside (to another network) will not know who pass it and therefore can not communicate with other networks.

USING THE PLATE / initial ARDUINO SHIELD ETHERNET Configuring Network
For an Arduino Ethernet (or Ethernet shield plate or some other
plate / shield that incorporates the same Wiznet W5100 chip) can start using
a TCP / IP network, the first thing you have to do is assign a range of settings (MAC address, IP address, etc.). To do write in the "setup ()" of our code the Ethernet.begin () function. This function has several ways of writing:

Ethernet.begin (Mac): where "mac" represents the MAC address you have our card. In the latest models of the shield Arduino Ethernet, this address is printed on a label, and therefore has to put that value, but on older shields you can choose any one. Any value of a MAC address is an array of type "byte" six items written in hexadecimal format, so normally declare and initialize the array previously in the area of global declarations (for example, as follows: Mimac byte [] = { 0xDE,
0xAD, 0xBE, 0xEF, 0xFE, 0xED};) and then assign within the "setup ()" that MAC address our plate / shield writing something
as this simply: Ethernet.begin (Mimac) ;.

This form of writing Ethernet.begin () function that is, just by specifying the MAC address and is- causes the plate / shield requested an IP via DHCP server to an existing local network. Therefore, he must have configured a computer within our LAN to act as a server

DHCP. If the connection to the DHCP server is successful, the Ethernet.begin () function returns a value of type "int" with value 1, and if not, return 0. All other forms of Ethernet.begin () explained below no return value.

Moreover, this form of writing Ethernet.begin () function specifying just the MAC address greatly increases the size of the final sketch compiled, an issue that we must consider not exceed the limit imposed by the microcontroller memory.

Ethernet.begin (mac, ip): where "ip" is the IP address (fixed, in this case) to manually attach to the plate / Arduino Ethernet shield. Any value of an IP address is an array of type "byte" four elements, written in decimal format, so normally declare and initialize the array previously in the area of global declarations (for example, as follows: meep byte []
= {10, 0, 0,
17}; ) And then assign within the "setup ()" that IP address

our plate / shield simply writing something like this: Ethernet.begin (Mimac, meep); He is assuming that "Mimac" has been declared the array containing the MAC-address.

There is another way to declare ip addresses instead of using a byte array type: IPAddress special declaration by meep (10,0,0,17); (Which in this example we call our ip "meep" and we have assigned the value 10.0.0.17).

Ethernet.begin (mac, ip, servdns): how to write this function is very similar to the previous one. Simply added to the plate / Arduino Ethernet shield a figure of additional configuration (in addition to its MAC address and fixed IP address) is the IP address of a DNS server that the board used. This IP address is an array of type "byte" four elements, written in decimal format, previously declared and initialized in the area of global declarations.

Ethernet.begin (mac, ip, servdns, gateway): how to write this function is very similar to the previous one. Simply added to the plate / Arduino Ethernet shield a figure of additional configuration (in addition to its MAC address, static IP address and the address of a default DNS server), which is the IP address of the gateway ("Gateway ") on our local network use this board. This IP address is an array type

"Byte" four elements, written in decimal format, previously declared and initialized in the area of global declarations.

In the forms of Ethernet.begin () where it is not explicitly specify the IP address of the gateway (the first three), this is assigned by default to the same IP Arduino own except in the latest issue of the four which is set to 1. That is, if the IP is 10.0.0.17 board decided that the IP gateway is automatically set, unless otherwise specified, to 10.0.0.1.

Ethernet.begin (mac, ip, servdns, gateway, subnet): this form of writing the function adds to the possibility of earlier forms also configure the network mask will have the plate / Arduino Ethernet shield. This data is an array of "byte" type of four elements, written in decimal format, previously declared and initialized in the area of global declarations. Its default value, if not explicitly specified, is 255.255.255.0

For that we used the first form of Ethernet.begin (), ie, which uses an external DHCP server for the IP of the plate / shield, it will be interesting to see two functions:

Ethernet.localIP () returns the IP address of the plate / shield. This return value must be previously declared type "IPAddress". If the IP is fixed you not make much sense to use this function, but in cases where this IP is dynamic, is the way to know exactly what ip obtained our plate / shield at a given time. No parameters.

Ethernet.maintain () allows the renewal of the concession of a dynamic ip. When an IP is assigned to a device by a DHCP server, this assignment is valid for a period of time determined by the server. With this function you can request a renewal in the allocation of ip granted. Depending on the server configuration, effectively renew the same ip, or a new different, or ip not be assigned extension is granted. This function has no parameters. Its return value is of type "byte" and can be: 0 (if nothing happens), 1 (if the granting of a new ip failed), 2 (if the granting of a new IP has been successfully completed) , 3 (if the renovation of the existing ip failed), 4 (if the renewal of existing ip has successfully completed).

The Ethernet library also offers the possibility of type UDP connections (using objects of type "EthernetUDP"), but not in this book
study. If you want to learn to use, I refer to the official documentation available on the website of Arduino.

Using as server Arduino

The TCP / IP networks typically have an architecture called "client-server". This means that the communication established between two machines, one of them takes the role of "client" and the other "server". The difference is in their behavior: what a customer is to make timely requests to a server, and what a server is to receive these requests and provide an adequate response to customer response. The server therefore has to be constantly "listening" to the possibility of receiving requests from customers, which may occur at any time and can come from multiple clients simultaneously.

The customer requests can be of many types: for example, a client can request the download of a file stored on the server, another client can request printing with a printer connected to the server, another client may request to see the web page that is hosted by the server, etc. In general, then, a server of a particular type provides a particular resource for customers of the same type. So, we will file servers and clients, servers and print clients, Web servers and clients, etc.

77

A single computer can have several types of server (web, printing, file, etc.). To prevent client requests from different types interfere with each other when connected with a "multi-type" server, a mechanism that is: each resource (service) offered by the server is "identified" by a number (called " port number"). Through a particular port, the server will only accept requests for a particular type of customers, so that a server can have multiple open ports, and each listening possible requests from one type of customer, without mixing. Ports are something like the windows of a bank: if we assume that the bank's customers represent requests, each must be targeted to the window to touch depending on the type of request it (if it is to make money You will have to go to a shop, whether it is for a mortgage to another, and so on); thus, requests for different types do not interfere with each other and everything works much more orderly and effective manner. It is very important that each client connects to the port that touch him, because if not, it is possible that communication is impossible between him and the server because it can not recognize that the request is coming.

When we say that the Arduino act as a server, we will be offering some kind of recourse permanently to any device capable of connecting to it (another board, a computer, etc.) can dispose of it. This means that our board should be open, as -a minimum port to listen and answer those requests received.

The first thing to do to turn our Arduino on a server is declared in the area of global declarations, variable (in fact, create an object) type "EthernetServer". This is done using the following syntax (assuming we call "myserver" to the variable): EthernetServer myserver (no port); where "port #" is an integer representing the port number used to listen to requests from clients connecting to the server.

Once already created the "myserver" object in the previous statement, which we implement by the following function:

miservidor.begin () causes "myserver" begins to listen to requests received through the port we have defined in the declaration. Normally, this function is performed within "setup ()" just after Ethernet.begin (). He has no parameters and return value.

78

From here, we can do things in our sketch. For example, we can send data to all customers (without distinction) that are already connected to our plate. To do this, we have three different functions:

miservidor.print (): Sends the data specified as a parameter to all clients currently connected to the plate-server. This data can be of any type: integer, decimal, or character string. If it is numeric, it is treated as a sequence of ASCII characters (ie, the number 123 is sent as three characters: '1', '2' and '3'). Optionally, it has a second parameter useful in the event that the data sent is complete, which may some value of the following predefined constants: BIN (to send the data in binary format), HEX (to send in hexadecimal) or DEC (to be sent in decimal format, even if the default is already well). Its return value is the number of bytes sent, but use is optional.

miservidor.println () works exactly like miservidor.print (), except that at the end of last shipment data as a parameter, adds ASCII characters 10 and 13, causing a line break.

miservidor.write (): Sends the data specified as a parameter to all clients currently connected to the plate-server. This data, however, can only be of type "char" or "byte". No return value.

But more interesting is deal individually with each of the connections that occur independently of and "customized". (Remember that a port of a plate / Ethernet shield acting as a server can support up to four simultaneously connected clients). To do this, we need the function:

miservidor.available (): Returns (creates) an object of type "EthernetClient" when "myserver" detects input data from any external client. This object represents the connection established with the client. If "myserver" receives no input, this function returns 0 and therefore no object is created. In any case, this object "EthernetClient" must be previously declared (in the area where you want our sketch) using the syntax: EthernetClient MyClient; (Assuming we call "MyClient" to that object). Once miservidor.available () has created the object "MyClient", the connection it represents can be manipulated by a number of characteristics of "MyClient" instructions pertaining only to that connection. In this way we can send or receive data communicating exclusively with "MyClient". Important to note that the managed by "MyClient" connection is persistent, and to explicitly close it is necessary to use micliente.stop () function. This function has no parameters.

After creating the object "MyClient" we can communicate with that particular customer in different ways. Specifically, to receive data from it (in fact, have been detected by miservidor.available ()) will use:

micliente.read (): returns a byte from "MyClient". Each time this function is executed, it returns the next byte received from that connection. If there are no more bytes available to read, it returns -1. This function has no parameters.

micliente.flush (): remove all the bytes that have reached the input buffer of the server from "MyClient" and not yet been read. He has no parameters and return value.

And to send data from "myserver" to "MyClient" We have several possibilities:
micliente.print () works exactly like miservidor.print () but this time only for connection "MyClient".

micliente.println () works exactly like miservidor.println () but this time only for connection "MyClient".

micliente.write () works exactly like miservidor.write () but this time only for connection "MyClient".

Example 8.1: Using a code example see everything clearer. To test this we need a sketch / Arduino Ethernet shield plate connected to a local network (typically through a network switch) and also one or more computers connected to the same network (which will make customer). The following code sends the data it receives through the open port (which is 23, but it could have been someone else) of a particular client to all clients (also including the sender) that are connected at that time. That also does the forwarding through the serial channel to the "Serial Monitor" (or equivalent), so if we see that data we received a USB cable connecting plate / Arduino Ethernet shield our computer.

```
/*Se necesita incluir la librería SPI porque el módulo Wiznet se comunica con la placa a través de este protocolo de comunicación. Nuestro sketch no hace uso explícitamente de las instrucciones de esta librería, pero las instrucciones de la librería Ethernet internamente sí. */
#include <SPI.h>
#include <Ethernet.h>
//La dirección MAC de nuestra placa/shield
byte mac[] = { 0xDE, 0xAD, 0xBE, 0xEF, 0xFE, 0xED };
//La dirección ip de nuestra placa/shield byte ip[] =
{ 192, 168, 1, 177 };
/*El puerto 23 es usado en servidores Telnet, pero
puede ser otro*/
```

81

```
EthernetServer miservidor(23);
/*Declaro el objeto " micliente" para poder
gestionarlo en el sketch, cuando se conecte*/
EthernetClient micliente;
void setup() {
        //Se inicializa la placa/shield
        Ethernet.begin(mac, ip);
        //La placa empieza a escuchar por el puerto 23
        miservidor.begin();


 Serial.begin(9600);
}
void loop() {
        int dato;
/*La primera línea mira se ha recibido bytes por el puerto 23. Si es así ,se crea el
objeto " micliente" correspondiente a esa conexión y se pasa a ejecutar el
interior del " if" . Si no es así, se vuelve al principio del loop() otra vez y continuar
mirando si se reciben bytes por el puerto 23, de forma infinita.*/
        micliente = miservidor.available();
        if (micliente > 0) {
/*Se lee byte a byte lo que llega del cliente " micliente" para reenviarlos a todos
los clientes conectados a " miservidor"  en ese momento y al canal serie de la
placa/shield*/
                dato=micliente.read();
                miservidor.write(dato);
                Serial.write(dato);

        }
}
```

To test the previous sketch we need a computer-client to connect to port 23 of the plate / shield Arduino and send some data. There are many programs that can do this, but the simplest is the Telnet client console, which usually included "factory" in most operating systems. To use either Windows or Linux, we must open the terminal command prompt and type telnet port number ipplaca. That is, in our case, telnet

192.168.1.177 23. From there, any written data, when pressing "enter" will be sent to the Telnet server (ie Arduino). It is also important to note that for computer-client correctly can make the connection within the local network must have an IP in the same network as the Arduino and the same mask. That is, it should have an IP type 192.168.1.x, where "x" is a number between 1 and 254 (other than the 177, which is the one with the Arduino board) and should have a mask as 255.255.255.0.

Besides Telnet client console, other programs could be used. For example, in Windows you can use a Telnet client (and more) with graphical configuration called Putty (http://www.putty.org). In Linux you can use the NetCat that can act as a substitute better able to Telnet command (http://netcat.sourceforge.net), versatile and flexible tool.

Using public ips to access Arduino

The previous exercise only works if our plate / shield and Arduino-client computers belong to the same LAN. But is not it possible to connect our plate / shield to the Internet, so that he could receive messages from any computer in the world? To do this, our plate / shield should have a public ip. We can achieve the same effect "linking" public ip that has the gateway of our network (our "router") to the plate / shield Arduino so that when the router receives a message is forward it to her. This is a procedure that has to be done by entering the router's control panel (usually via web), and varies from model to model. However, we do not study because it has a drawback: most of the time, the public IPs assigned by the telephone operators are changing from time to time according to their criteria, so our public ip router one day may be a and another day may be another. This means that computers-customer should know at all times what public ip corresponds to our plate / Arduino shield. And this is not very practical.

A simple way to avoid this complication is to use a dynamic DNS service (DDNS), as offered by http://www.no-ip.com or similar. The idea is to link the public IP of our router with a DNS name chosen by us, so-client computers do not have to know our public ip to connect, but it is enough to use that DNS name. What matters is that this service will automatically update the public ip link <-> DNS name whenever our public ip change, so do not worry we will never have this problem. So, using a dynamic DNS server, our gateway will have an accessible name for all computers in the world.

How to specifically use the service non-IP is very simple. First we create a user account and then we loguearnos with it on its website. Once there we will have the "Add Host" option, which will lead to a form where you can choose the DNS name you want for your router. A name that will always add their own "tagline" No-Ip, we can also choose from a number that is in a drop box. For example, a name might be "mirouter.no-ip.org" or "mirouter.zapto.org" depending on the selected tagline. In that same form make sure the "DNS Host (A)" which already defaults, and nothing is checked. The ip shown us in that form to the public IP of our router detected corresponds; the hand we can say if we know and we know that shown is incorrect, but this is very

likely. After pressing the button "Create host" form, we still have to do more to get our router available on the Internet with the name chosen must configure your router to make use of non-IP service.


The way to achieve this varies depending on the router model, but in any case it is an existing option in the control panel of the device, which is normally accessed via web specifying the private IP in a browser and a username and password in the box that appears. These three data has to provide the manufacturer. Anyway, in the section on configuration DDNS we will ask the company offering the service (in our case, No-IP), and the username and password used in creating the DNS name. Having done this, the router will maintain communication with No-IP to automatically notify a change in public ip link <-> name, so that, now, our router is always available on the Internet. Note: There may be some model of router does not have the list of services the Non-configurable IP DDNS service; in that case, or it has to use a similar service that supports itself or, you can try to overwrite the firmware of the router by another that support No-IP, such as DD-WRT (http: // www .dd-wrt.com). However, the latter solution is only recommended for users who really know what they're doing, because you can end up with a useless router. 85

However, there is still a step further: using our non-IP router is already accessible online, but within our LAN we can have multiple computers connected to it (including our plate / Arduino shield). How does the router when you receive a message, addressed to a particular computer from which you are connected? By redirecting ports. This is another section of the control panel of the router, usually marked "Port Forwarding", which can forward messages that the router will reach that target a specific port to a specific port on a specific host on the LAN, identified by private ip. That is, if we assume that our plate / shield Arduino has port 23 open (as in the example above), in setting up port forwarding on the router would have to specify that messages received by the port 23 of the router to be redirected to port 23 machine our LAN with private IP 192.168.1.177 (in our example). And now that we will have our plate / shield open to hear requests from around the world, literally. Keep in mind, however, that each port can only be redirected once.

Free services Non-IP expire in case of inactivity: If the DNS name is not accessed within 30 days, the domain will be deleted from the system. Is expires possible to prevent the service by clicking on a link in an e-mail expiration warning is sent after 25 days of inactivity, or even buying a No-IP service payment.

## Use as client Arduino

If we want our Arduino to connect as a client to any other network device that acts as a server (ie, our Arduino request an external resource), the first thing to do is declare an object of type "EthernetClient "(in the context of our sketch where desired), representing the board itself. Assuming this EthernetClient object we call "MyClient", the next thing to do is connect to the server you want. This is done with the following statement:

micliente.connect (): connecting with the server whose IP is specified as the first parameter (you can also specify a DNS name, if a DNS server previously was written in Ethernet.begin ()) and whose port number has been specified as the second parameter. If an IP is specified, this must be an array of "byte" type of four elements previously declared and initialized. If a DNS name, this is simply a string is specified. The return value of this function is "true" if the connection is successful or "false" otherwise.

It is clear from the above function to make a connection to an external server, our plate / Arduino shield must, in addition to its ip, make appropriate port number. Any one not worth writing, specify the specific port number that the server supports requests, and all requests to a different port that provides the server will be ignored. Luckily, there are a number of standardized port numbers that usually always used for the same tasks; thus, for example, the HTTP server (also known as server "web", ie, computers that provide web pages for clients HTTP-the so-called "browsers" - to visit them) typically use port 80. Therefore, if our Arduino be connected to an HTTP server, you must write your ip and port 80 in micliente.conect ().

87

Other examples of servers are FTP servers (servers that allow Ascent bajada- file transfer and FTP clients), which have

O port 21. Open SSH (servers that allow remote access from client SSH terminal commands) servers, which have opened the 22. There are also Telnet servers (similar to those SSH but less secure, since not encrypted the information transmitted), which open port 23. And the SMTP servers (used by SMTP clients to send emails to your destination) and POP3 server (used by POP3 clients to read emails from our mailbox) which use port 25 and 110, respectively. And so to thousands of different servers. If you want to know the full standard list, you can consult http://www.iana.org/assignments/service-names-port-numbers/service-names-port-numbers.txt.

To send data from "MyClient" the server to which we are connected, you can use the functions:

micliente.print (): their behavior is exactly the same as micliente.print () seen in the previous section, but this time the transmission of data from the Arduino board is acting as a client to an external server.

micliente.println (): their behavior is exactly the same as micliente.println () seen in the previous section, but this time the transmission of data from the Arduino board is acting as a client to an external server.

micliente.write (): their behavior is exactly the same as micliente.write () seen in the previous section, but this time the transmission of data from the Arduino board is acting as a client to an external server.

88

To receive data from the server (typically a response to the request sent by the client), we can use:

micliente.available (): returns the number of bytes available to read (it returns 0 if there is none, of course) from the server. That is, the amount of data being sent from the server to "MyClient". No parameters.

micliente.read (): their behavior is exactly the same as micliente.read () seen in the previous section, but this time the data is received by the Arduino board acting as a client from an external server.

micliente.flush (): remove all the bytes that have reached the input buffer "MyClient" from the external server that has not been read. He has no parameters and return value.

Finally, we have two control functions more:

micliente.stop (): off "MyClient" the external server.

micliente.connected () Returns "true" if it is "MyClient" connected or "false" otherwise. Often it is the server that closes the connection, and this function serves to detect whether this has occurred. In that case it's natural that connection also close by us, by micliente.stop (). Note that it is considered that a client is connected but the connection is closed if there are still unread data. This function has no parameters.

Example 8.2: Using a code example see everything clearer. In this sketch, our plate / Arduino shield connects to the Google search engine (and therefore to a web server listening on port 80), and sends you a particular character string representing a search request, namely web pages containing the word "Arduino". We can see by the "Serial monitor" the answer the search engine Google brings us. For this example to work (not very practical, but very illustrative), the plate need only have access to a door functional link in our LAN (usually through connection to a network switch) with ip 192.168.1.1 . The board must also be connected via USB to our computer.

```
#include <Ethernet.h>
#include <SPI.h>
byte mac[] = { 0xDE, 0xAD, 0xBE, 0xEF, 0xFE, 0xED };
byte ip[] = { 192, 168, 1, 177 };
//Una ip del servidor web de Google (en concreto, su buscador)
byte  miservidor[]  =  {  173,  194,  67,  103  };
EthernetClient micliente;
void  setup()  {  Ethernet.begin(mac,
        ip); Serial.begin(9600);
        /* Si micliente.connect() devuelve 0, la conexión no se ha podido
        realizar. Los servidores web siempre escuchan en el puerto 80 */
        if   (micliente.connect(miservidor,   80)   !=   0)   {
            Serial.println("Conectado");
```

```
                    /*Se envía esta cadena de caracteres al buscador de
                    Google.Veremos su significado enseguida */ micliente.println("GET
                        /search?q=arduino      HTTP/1.1");      micliente.println("Host:
                        www.google.com");
                        micliente.println();//Línea en blanco:marca de final
                }
        }
        void loop(){
                    char c;
                    /*Si hay datos disponibles que han llegado desde
                    el servidor pendientes de leer*/
                    if (micliente.available() > 0) {
                            //...se lee un byte en cada repetición del loop c =
                            micliente.read();
                            /*...y se muestra ese byte por el " Serial monitor" . El resultado
                            final será     la respuesta del servidor Google a la cadena que se
                            ha enviado en setup()*/
                            Serial.print(c);
                    }
                    //Si el servidor de Google ha cerrado la conexión if
                    (micliente.connected() == 0) {
                            micliente.stop();   //La cerramos nosotros también
                            for (;;){;} //Y no hago nada nunca más
                    }
        }
```

What does the string "GET / search? Q = arduino HTTP / 1.0" in the previous code sent to Google search engine? And the "Host: www.google.com" line? We have said that what we did was a request for web pages that contain the word "Arduino", but why has this syntax? Because it is a HTTP request.

Brief note on the HTTP protocol:

The HTTP protocol is a basic language consisting of questions and answers that all web clients (ie browsers) and web servers (the programs offered by web pages worldwide) understand. Although using different web clients (Firefox, Chrome, Safari, Internet Explorer ... or in this case, the Arduino itself) and "the other side" is running different Web server software (Apache, Nginx, IIS, etc.) the HTTP protocol is standard and enables communication between both ends of a universal form.

Basically, the HTTP protocol consists of specific requests that can run a Web client, and predefined answers that can offer a web server to these requests. Among the types, the most common potential applications far is to request a web page; this request is done by sending the "GET" command to the server; and after him, the name of the specific page you want to get on that web server.

In the code example above, the symbol "/" means that the requested page is the homepage of the website. And the chain that follows (search? Q = Arduino) is the parameter that is passed to this home page to be interactive (in this case, to look for the word "Arduino"). In fact, we can see easily how, if we access the page of the Google search engine through a normal browser, depending on what we write in the text box and the tail "search? Q =" is amended visible address in the address bar. Finally, HTTP / 1.1 string specifies the protocol version that the web client is using (in this case, the 1.1, which is the latest); this chain has always put at the end of the GET command.

If we wanted to make a GET request to a different page other than the main site (eg http://arduino.cc/en/Reference/HomePage instead of http://arduino.cc), we would have add after the slash "/" across the chain after the address of the web server. That is, in the example above should be GET request GET / in / Reference / homepage HTTP / 1.1

Normally, HTTP requests are not composed only of a GET line but are composed of more lines, if not specified, they take a default value. These lines (collectively called "the head" of the client application) are used to inform the server about more technical details of the request. Where xxxx always represents the DNS name of the Web server that the client wants to connect: In the above code specific header line, which is the "xxxx Host" line is sent. This line is essential to state it if the version is used HTTP 1.1, so I usually always see a GET request accompanied by a header line "Host: xxxx".

HTTP replies from the server after a request by the customer can be varied: If the server can successfully provide what the customer requested, the server suggests sending a response code such as "HTTP / 1.1 200

93

OK ", where that string is the protocol version used and a standard preset numerical code that identifies the meaning of the response (and a text more understandable) it specifies. Each type of response has a numeric code indicating the predetermined type of response. For example, the famous 404 that appears when accessing a non-existent page is precisely the HTTP code that the server sends the client informed of this fact. For the set of numerical codes server response, we can consult the official document of the HTTP standard http://www.ietf.org/rfc/rfc2616.txt.

Also happens in client requests, responses from servers not only consist of the line with the ID code type of response and that's it, but are more specific lines formed by the HTTP protocol. These lines (collectively referred to as "the head" of the response from the server) serve to inform the customer about the answer more technical details.

Another question that we have running the above code is the meaning of what we visualize the "Serial Monitor". We have said that is the answer that Google offers based on the request that we have done, but what has that answer? It consists of two parts: the first lines are the head of the server response. And after an empty line (blank), starts sending itself the content requested by the client (ie the web page). In this case it would be the first website to the list of results offered by Google after clicking the button "Search". Logically, if the application is conducted through a browser, what we see is that page, but if the application is conducted through the plate / Arduino shield, the visible results by the "Serial Monitor" no resemblance. Why? Because what we are actually seeing is the HTML code of the page.

Brief note on the language of HTML tags:

All web pages are written basically in a "language" called HTML. When a browser requests a web page, which is precisely the server receives the HTML source code (which is what we see in the "Serial Monitor". The function of the browser is able to interpret the HTML code into visible elements for the user ( images, paragraphs, titles, etc.). Because the HTML language is standard, no matter which browser we use in principle, because the page should be displayed correctly.

HTML is not a programming language really, but a simple set of tags that indicate where you need to display a picture or text; what it is called a "markup language". The study is beyond the aims of the book
This standard, but if you know you want to start, I recommend the great tutorials and www.w3schools.com/html5 http://www.w3schools.com/html present.

Example 8.3: We can modify the previous code so that rather than always performing the same search (the word "Arduino"), we can decide each time the execution of the sentence sketch Google search starts. To do this, we introduce the phrase using the "Serial Monitor".

```
#include <Ethernet.h>
#include <SPI.h>
byte mac[] = { 0xDE, 0xAD, 0xBE, 0xEF, 0xFE, 0xED };
byte ip[] = { 192, 168, 1, 177 };
byte servdns[] = { 8, 8, 8, 8 };
//Usamos su nombre DNS en vez de su ip
char servidor[] = "www.google.com";
EthernetClient micliente;
//Frase a buscar
String frase; byte
i=0;
void setup() {
        char c;
        Ethernet.begin(mac, ip, servdns);
        Serial.begin(9600);
          if (micliente.connect(servidor, 80) != 0) {
                Serial.println("Conectado");
//Mientras no se escriba la frase a buscar, no se hace nada while
(Serial.available() == 0) {;}
/*Arduino lee carácter a carácter lo que se ha escrito en el
"Serial monitor" y los guarda en un String */
                while (Serial.available() > 0){
/*Es importante que " c" sea de tipo " char" y no " byte" para que en
la frase no se guarde el código numérico ASCII sino el carácter en sí*/
                        c=Serial.read();
//Añadimos el carácter leído al final de la frase
                        frase.concat(c);
```

/*Este delay es muy importante para darle tiempo a Serial.available()

a recibir el siguiente carácter en el buffer. Si no se pone, el código

Arduino es más rápido que la llegada de nuevos bytes y la función Serial.available()

puede devolver 0 cuando todavía no se han recibido todos los bytes */


}
} Else {


delay (100);

Serial.println ("not connected");

}

if (micliente.connected ()! = 0) {micliente.print ("GET / search? q =?"); micliente.print

(phrase); micliente.println ("HTTP / 1.1");

micliente.println ("Host: www.google.com");

micliente.println ();

}

}

void loop () {

char c;

// If I receive server response ...

if (micliente.available ()> 0) {

// The show character to character in the "Serial Monitor" micliente.read c = ();

Serial.print (c);

}

```
if (micliente.connected () == 0) {Serial.println ("Off");

micliente.stop ();

for (;;) {;}

}

}
```

Case study: on-board / shield Arduino web server

In the following code, we turn our plate / shield into a single web server. That is, we make our plate / shield can answer HTTP requests to provide customers with HTML content. Therefore, from a computer either could open a browser and type in your address bar private IP plate / shield (if we are in a LAN) or DNS name (if connected through Internet) so contact she and display the HTML content of this website as. In this case, we assume that we have connected to the plate / shield any analog sensor (an LDR, a thermistor, etc.) whose readings will be visible on the website if we have no sensor by hand, the code will work perfectly, only the read values will be random noise.

```
#include <SPI.h>
#include <Ethernet.h>
mac byte [] = {0xDE, 0xAD, 0xBE, 0xEF, 0xFE, 0xED}; IPAddress ip
(192,168,1,177);
EthernetServer myserver (80); EthernetClient MyClient;
// Used to know when the application has just received from the client
lineaActualEstaVacia boolean = true;
/ * Each character received from the customer as part of the HTTP * / char character
request;
/ * Pin Arduino where it is assumed that the analog sensor * / byte pinAnalogico = 0
is connected;
```

```
// Returns reading reading analog sensor int = 0;

void setup () {Ethernet.begin (mac, ip); miservidor.begin ();

/ * To view the request sent from client to server (request headers) * /

Serial.begin (9600);

}

void loop () {

miservidor.available MyClient = ();

// If a connection of a client if (MyClient == true) is detected {

Serial.println ("New client connection");

while (micliente.connected () == true) {

// Proceed to read character at the HTTP request

if (micliente.available ()> 0) {

micliente.read character = ();

// For the "Serial monitor" I see the header lines of the petition

Serial.print (character);

/ * If the character received is a new \ n (newline) it means that we have reached the
end of that line. And if that lineaActualEstaVacia is true, it means that the client has
sent us a completely empty (blank) entire line without characters. A completely
empty line is the standard signal to indicate that the HTTP request

It has ended, and therefore, can send a response. * /

if (character == '\ n' && lineaActualEstaVacia == true) {

/ * Send a standard HTTP header. This consists of the line with the response code
200 (OK), another line
```

called Content-Type: customer indicating the type of website that you are going to send (usually, it is always "text / html") and called line Connection :, whose value "close" indicates that the server will close the connection with the client once sent the requested web page (ie, do not make a persistent connection) * / micliente.println ("HTTP / 1.1 200 OK."); micliente.println ("Content-Type: text / html"); micliente.println ("Connnection: close");

/ * A blank line marks the end of sending HTTP headers. From here the HTML content itself is sent. In this example the Arduino sends a single HTML page, so this is the default web page. This means no special indication is in the browser to view it apart from specifying the ip / name plate. * /

micliente.println ();

/ * The HTML code consists of a series of "tags" that indicate the type of content that can be displayed. These labels are indicated by "<" and ">". The next two lines are two labels which necessarily must always appear at the beginning of the HTML code of any page * /

micliente.println ("<! DOCTYPE html>");

micliente.println ("<html>");

/ * This tag tells the browser to refresh the page (that is, to re-make a request to the server) automatically every 5 seconds. If we do not add this line, the browser only get once the read data from

sensor, and to get it every time we make a new request "by hand" * /

micliente.println ("<meta http-equiv = \" refresh \ "content = \" 5 \ ">");

/ * Pick up the value of the analog sensor and

show within the HTML code that will see the browser * / read = analogRead (pinAnalogico); micliente.print (reading);

/ * The label "<br />" serves to add

a jump of line visible on a web page. * /

micliente.println ("<br />");

/ * Any HTML code must end with the "</ html>", which also marks the end of the content sent by the server * /

micliente.println ("</ html>");

/ * There's nothing more to send. Out of the loop and either close the connection * /

break;

}

// Just a line and start a new one, in principle empty if (character == '\ n') {

lineaActualEstaVacia = true;

/ * The read character is any one of that line, so it is not empty. Except for the character \ r (carriage return) comes from that \ r always precedes \ n in all instances, so in reality, an empty line always has the character \ r. So in the case of detecting the character \ r, the line is still considered empty. That is basically what you're looking for is a sequence like this:

line characters, including \ r -> lineaActualEstaVacia = false

\ N (newline) -> lineaActualEstaVacia = true

\ R (character "Ghost", nothing happens) -> lineaActualEstaVacia = true

\ N (newline) + lineaActualEstaVacia = true -> Shipping response * /

} Else if (character! = '\ R') {

lineaActualEstaVacia = false;

}

}

}

```
// We time the browser to receive the delay data (1);

/ * As we have already sent the page to the client, we have nothing more to do with
him, so close the connection * / micliente.stop ();

}

}
```

**Case Study: SD card web server**

Suppose you have a file named "page.html" stored in an SD card connected to our plate / Ethernet shield. In the previous example we saw how to serve a web page generated in real time, but

How can we serve a web page that is previously stored in the SD card? Well, in a very similar way than we have seen. Following is a sample code where the only change over the previous code is the response that the server provides the client after sending the appropriate HTTP headers are presented.

```
#include <SPI.h>
#include <Ethernet.h>
#include <SD.h>
mac byte [] = {0xDE, 0xAD, 0xBE, 0xEF, 0xFE, 0xED};
ip byte [] = {192,168,1, 177}; File page;
EthernetServer myserver (80); EthernetClient MyClient;
char character;
lineaActualEstaVacia boolean = true;
void setup () {Ethernet.begin (mac, ip); miservidor.begin ();
pinMode (10, OUTPUT); Bookseller // Required for SD
/ * If an error occurred in initializing the card, the program is aborted * /
if (SD.begin (4)) {return; }
}
```

```
void loop () {

miservidor.available MyClient = ();

if (MyClient == true) {

while (micliente.connected () == true) {

if (micliente.available ()> 0) {

micliente.read character = ();

if (c == '\ n' && lineaActualEstaVacia == true) {client.println ("HTTP / 1.1 200 OK");

client.println ("Content-Type: text / html"); client.println ();

/ * After sending HTTP headers the file stored on the SD card and reads character at

opens. While each character is read, this is forwarded by the network to the client. * /

page = SD.open ("page.html");

if (page! = 0) {// If there is no error ...

// While characters in the file for reading ...

while (pagina.available ()> 0) {

// ... The forwarding to the client micliente.write (pagina.read ());

}

// After all characters read, I close the file pagina.close ();

}

break; // I've sent everything. I go to close the connection

}
```

```
if (character == '\ n') {
lineaActualEstaVacia = true;
} Else if (character! = '\ R') {
lineaActualEstaVacia = false;
}
}
}
delay (1);
micliente.stop ();
}
}
```

The above code is usually used primarily to display web pages whose content consists of information obtained from a sensor periodically and recorded in real time on the SD card. As an example of code where you see how they are stored on an SD card data received by a sensor we can recall the studied in this book under the heading of humidity sensors, so do not recommend rereading. In this example, the data is stored in simple text files, but in the case of web pages want them saved in the process is very similar, because in reality, the web pages are just text files written using certain rules ( more specifically, using a certain HTML) tags. Therefore, if we write those labels properly, we can store information so we generate a standard web page (viewable by any browser), instead of generating a simple tab-delimited text file.

The previous code also we could have added the ability to offer different files under certain circumstances (the push of a button, reading received from a sensor, etc.). It is left as an exercise.

Another option is left as an exercise to the user connected to our plate / shield Ethernet viewing a web page that displays a menu of links, each corresponding to a stored file on the SD card; in this way, the user could choose which file each time you want to download. The code allows for this possibility is certainly more complex and lengthy than we are showing in this book, but no less interesting study. What can we get from here: https://github.com/adafruit/SDWebBrowse.

## Case study: web form control actuators

In the example in the previous section, we assumed that we had a number of sensors connected to our plate / Arduino shield and we wanted access to the values read by them through a website offered by the board itself. But what if we want to use that site not to observe sensor data but to control the behavior of actuators? That is, to move motors or LEDs light touch of a button visible within our website. What should we do?

For something very similar to the previous example we have set up a web server on the plate / Arduino shield that accepts requests from client computers, which will see a simple website hosted by her. What changes now basically is the HTML code for that page, because instead of showing a value and is now (as did in the example above), this time the page has to display a series of buttons that have to react in a certain way when pressed (moving a motor in one direction or another, speeding it up or slowing it down, turning on or off one LED, emitting a melody with a buzzer or other, etc.).

The following code is offered as reference template from which you can expand and extend its functionality to introduce new actuators. As presented, moves a servomotor connected to the PWM output pin No. 6 in three ways (by three buttons). Each time you press the first button the actuator will move 20 degrees in one direction, each time you press the second button moves 20 degrees in the other direction, and each time you press the third servomotor will move from 0 to 180 degrees, and vice versa.

```
#include <SPI.h>
#include <Ethernet.h>
#include <Servo.h>
mac byte [] = {0xDE, 0xAD, 0xBE, 0xEF, 0xFE, 0xED}; IPAddress ip
(192,168,1,177);
EthernetServer myserver (80); EthernetClient MyClient;
Servo miservo;
int pos = 90; // Is not "byte" because "byte" does not support negative boolean
lineaActualEstaVacia = true;
char character; String request = ""; void setup () {
```

```
Ethernet.begin (mac, ip); miservidor.begin (); miservo.attach (6);

}

void loop () {

int i;

miservidor.available MyClient = ();

if (MyClient == true) {

while (micliente.connected () == true) {

if (micliente.available ()> 0) {

micliente.read character = ();

/ * We only want to store the first character of the request to find out which button is

pressed (we have set 30 for convenience, but you can change); we do not care to

keep all the complete request * /

if (peticion.length () <30) {

peticion.concat (character);

}

if (character == '\ n' && lineaActualEstaVacia == true) {micliente.println ("HTTP / 1.1

200   OK");  micliente.println  ("Content-Type:   text  /  html");  micliente.println

("Connnection: close"); micliente.println ("");

micliente.println ("<! DOCTYPE html>");

micliente.println ("<html>");

/ * All there is between the label

<H1> and </ h1> is a title * /

micliente.println ("<h1> Form servo control </ h1>");
```

/ * Here we check what has made the request client (that is, which button was pressed browser). Depending on the option chosen, you will be shown on the website a phrase or another and the servo will move accordingly. This area code can be extended to what you want: an LED light, sound a buzzer, etc * /.

```
if (peticion.indexOf ("SentidoServo = 0")> 0) {

micliente.println ("Sense of the current movement: <br/> Left");

micliente.println ("current angular position");

pos = pos + 20;

/ * If within if there is only one instruction can be written without braces * /

if (pos> 180) pos = 180;

miservo.write (pos);

micliente.println (pos);

}

if (peticion.indexOf ("SentidoServo = 1")> 0) {

micliente.println ("Sense of the current movement: <br/> Right");

micliente.println ("current angular position");

pos = pos-20;

if (pos <0) pos = 0;

miservo.write (pos);

micliente.println (pos);

}

if (peticion.indexOf ("SentidoServo = 2")> 0) {

micliente.println ("Sense of the current movement <br/> Lap");

for (i = 0; i <180; i ++) {

miservo.write (i); delay (5);

}
```

```
for (i = 180; i> 0; i -) {

miservo.write (i); delay (5);

}

}

/ * Once the checks made control servos, I just sent the rest of the page is missing.
This consists of a form with the control buttons * /

micliente.println ("<br/>"); micliente.println ("<form>"); micliente.println ("<button type
= 'submit'

name = 'SentidoServo' value = '0'> Left </ button> ");

micliente.println ("<button type =" submit "name = 'SentidoServo' value = '1'> Right
</ button>");

micliente.println ("<button type =" submit "name = 'SentidoServo" value = "2"> Lap </
button> ");

micliente.println ("</ form>");

micliente.println ("</ html>");

break;

}

if (character == '\ n') {

lineaActualEstaVacia = true;

} Else if (character! = '\ R') {

lineaActualEstaVacia = false;

}

}

}
```

```
/ * Empty string current client request to start again at the next possible connection *
/
request = "";
micliente.stop ();
}
}
```

Most important to fully understand the above code is how the mechanism works sending requests from the browser to the Arduino. Actually, we have used the simplest way possible to interact via the web with the Arduino: using buttons in an HTML form. Thus, if a button is pressed, a specific request is sent, if another is pressed, it sends another, and plate responds to each request accordingly.

An HTML form always starts with the <form> and ends with the </ form>. Between these two labels have to place the elements that will form the form, which can be of many types (text boxes, combo boxes, "checkboxes", etc.). To us the form element that interests us is the submit button data. For each button you want to include this type, we write a <button> with a series of "benchmarks". The parameter "type" is used to indicate that the function of this button is to send a piece of information to a web page for this processing. That website must be defined within the <form> via the "action", but if this is not specified (as we have done), then the data is intended for the same website where emerged. This is what interests us because the Arduino board shows only one page that performs two things: show the form and process data sent from it. The parameter "name" button serves to indicate the name of the data to be sent, and the parameter "value" is used to indicate its value. <-> value, so you can easily identify what the data to be processed, and what its precise value at that time That is, the data are always sent in pairs name. Thus, for example we could add to our code other buttons for controlling an LED, or a DC motor, or a buzzer, etc., they had another name to distinguish them from those who already control the servomotor, and make each one of These new buttons send a different value (HIGH / LOW, or an analog value).

Well, we know how the form sends the data. But how do you process? Unless otherwise specified, the default type of sending that data is using the "GET" system (though you can also use other shipping something different system, called "POST"). Shipping type "GET" means that in the first line header request the name pair appears <-> data value

sent using the following syntax: GET / name = HTTP / 1.1 value. Therefore, a simple way to check if you have pressed a particular button is to detect if the "name = value" Searches on the previous chain. The Arduino language has String objects for instruction "indexOf ()", which can serve us for this, and remember that this command always returns a value greater than -1 if the search string is within another major.

Keep in mind that we have restricted to 30 characters (this is for convenience, can be changed if needed) the string used to search within it the data "name = value" sent. We could have added the entire header of the request within that chain, but in that case we could have encountered a problem. Besides being in the first line, the data "name = value" may appear in other lines of the header (such as "Referer:" line), so in that case the data would be detected several times and therefore the servomotor's only really necessary steps would move. To avoid this problem, to 30 characters (a safe length) we fill the string used to search within it the data sent, so we got that contains only as much once the data "name = value" without repeating .

You can build more complex HTML forms that are able to send multiple values at once in a GET request. In such cases, the syntax of the request is GET /? Name1 = value1 & name2 = value2 HTTP / 1.1 (for sending two data) or GET /? Name1 = value1 & name2 = value2 & Name3 = value3 HTTP / 1.1 (for sending three data ), and so. In any case, this is easy to check if you look at the address bar of your browser: there is written the name couples particular string <-> value sent after clicking the button. In fact, it would be completely equivalent to write the name string <-> desired value directly into the address bar and press "Enter" to press the appropriate button.

Concerning the design of the websites offered by Arduino, is clearly not the most advanced in the world. If we want to improve the appearance of these, it is best to use a "language" called CSS HTML complementary, which is specifically designed to control the aesthetic appearance of the web pages. Combine CSS with HTML display makes this a much more colorful way. However, it is beyond the scope of this book to delve into this subject. If you want to know more, I recommend consulting the excellent online tutorial available in http://www.w3schools.com/css and http://www.w3schools.com/css3.

To conclude this section, we will say that instead of using buttons in the form to control the servo state, we could have used other
data entry elements, such as a box. If we substitute in the above example code lines

micliente.println ("<button type =" submit "name = 'SentidoServo' value = '0'> Left </ button>");

micliente.println ("<button type =" submit "name = 'SentidoServo' value = '1'> Right </ button>");

micliente.println ("<button type =" submit "name = 'SentidoServo" value = "2"> Lap </ button> ");

these:

micliente.println ("<input type = 'text' name = 'SentidoServo' />");

micliente.println ("<button type =" submit "value =" 0 "> Submit </ button>");

obtain a form with a text box and a button. In the box we enter the control value ("0" in one sense--movement, "1"

in another sense-or-movement "2" -Return completely) we want, and the button will help us to send it to Arduino. Any value other than "0", "1" or "2" does nothing. You have to look in the corresponding text box HTML element is the <input /> element which must have at least two parameters: type (standard text boxes are of type "text") and its name, that will serve the Arduino to identify which box the data received belongs. Obviously, a text box does not have parameter "value" explicit because this is the text that has been introduced at all times. The <button>, meanwhile, is identical to those already seen, except that now you need not send any "value" because its only function is to send the data entered into other form elements.

From the above examples, we can be able to build a web control panel to manage any element connected to an Arduino board. For example, we could use text boxes to set the message to be displayed on an LCD, or to vary the speed of a DC motor, or to choose the melody to be issued by a speaker, etc.

Case study: sending messages to Twitter.com

Twitter (http://www.twitter.com) is a free microblogging platform. That is, after registration, you can create a message of maximum 140 characters and make it visible to everyone (literally), but for other Twitter users can receive these messages automatically,

must subscribe to them. The way to read the messages received through our subscriptions vary: accessible (after entering username and password) through the official website Twitter, or using a specific application for PC or mobile phone of last generation.

We can use this simple operation to make our Arduino can post messages using its own Twitter account and then use our own personal account to subscribe to those messages. Thus, at all times we will be informed of everything that our Arduino notify (when some event or at regular intervals or when we want is detected). If we care about the privacy of these messages issued by the board, you can always set up your account so that only specified users can read.

After creating the account in http://www.twitter.com/signup our Arduino (and one for our use), to which the plate can publish we must unburden the library "Arduino Tweet" messages here: http://arduino-tweet.appspot.com. Warning: This library does not send messages to Twitter.com but sends his own site, and from there forwarded to Twitter.com. That is, a web of third parties is used as an intermediary. This is because as Twitter.com has a relatively complex called OAuth authentication mechanism, the library "Arduino Tweet" chooses to delegate to an auxiliary web all the necessary credentials exchange process of granting access to an account, rather than falling upon the Arduino itself. In this way, we gain ease of use and not saturate our work Arduino. But instead we rely on a third party service, which already informs us that does not accept more than one shipment of messages per minute (not supersaturated). If you do not want to use this system and instead wants to lie around the OAuth process Arduino own to avoid intermediaries, there is a library with support for OAuth Arduino (http://www.markkurossi.com/ArduinoTwitter), but certainly , installation and use is much more complex.

Once we installed the library "Arduino Tweet" before it can be used there to take a first step. You have to get a "token". A "token" is nothing more than a set of characters that will be used to authorize the web http: // arduino- tweet.appspot.com access to our Twitter account Arduino. The "token" what we write in our sketch to work. To achieve this we must first loguearnos on Twitter with our Arduino account and then click on the marked "Step 1" of the web http://arduino-tweet.appspot.com and follow the steps that appear bond.

On the issue of security using the library "Arduino Tweet" not to worry, because the intermediate web is not able to know the username and password of the Twitter account of our Arduino: OAuth mechanism prevents own . In any case, if after testing do not want http://arduino-tweet.appspot.com can still access the account Arduino, we can always take that privilege logging in with that account and going to the Control Panel http : //twitter.com/settings/connections.

The following code sends a message to Twitter (always the same) each time you restart the Arduino board. Keep in mind, however, that rejects equal consecutive Twitter messages (error 403), so surely, despite running this code several times, just leave the message the first time if we do not change the text.

```
#include <SPI.h>
#include <Ethernet.h>
#include <Twitter.h>
mac byte [] = {0xDE, 0xAD, 0xBE, 0xEF, 0xFE, 0xED};
ip byte [] = {} 192,168,1,177;
servdns byte [] = {8,8,8,8}
gateway byte [] = {} 192.168.1.1;
subnet byte [] = {255, 255, 255, 0};
// Xxxx must be replaced with the token of the Arduino account
Twitter twitter ("xxxx");
void setup () {
Ethernet.begin (mac, ip, servdns, gateway, subnet); Serial.begin (9600);
tweet ("Hello");
}
```

```
void loop () {}

tweet void (char msg []) {boolean sent = false; int codigorespuesta;

// I connect to Twitter to send the message sent = twitter.post (msg);

// If you've contacted Twitter ...

if (sent == true) {

/ * ... I wait until the message has been sent successfully or an error has occurred. In
any case, I get the answer as shipping HTTP server header code. The "& Serial"
parameter is to relay the response to the serial channel; if that is not needed, is just
twitter.wait write (); * /

twitter.wait codigorespuesta = (& Serial);


if (codigorespuesta == 200) {Serial.println ("OK.");

} Else {

Serial.print ("Something went wrong"); Serial.println (status);

}

} Else {

Serial.println ("Failed to contact Twitter.");

}

}
```

From here it is easy to write a skit to react to various events and send corresponding messages. To simulate these events, we assume that we have for example two buttons: pressing one posts a message and pressing the other a different message will be sent. The only thing we should then modify the code above is the "setup ()" to indicate the input signal pins, and the "loop ()" as shown below: All other code will remain intact.

```
void setup () {
Ethernet.begin (MAC, IP, gateway, subnet); Serial.begin (9600);
pinMode (6, INPUT);
pinMode (7, INPUT);
}
void loop () {
msg1 char [] = "occurred event 1"; msg2 char [] = "event occurred 2"; if (digitalRead
(6) == HIGH) {
tweet (msg1);
delay (2000); // We time the sending and receiving
}
if (digitalRead (7) == HIGH) {
tweet (msg2);
delay (2000);
}
}
```

**Case study: Sending data to Cosm.com**

Cosm.com (http://www.cosm.com), formerly known as Pachube, it is a platform that allows different applications (written in different languages, such as Java, C, Processing ...) connected to a data warehouse also online

of a presentation graphics system online. Using appropriate procedures, these applications can store information in Cosm.com time to time and view it in the form of statistical graphs. In practice, this platform is typically used to monitor sensors geo systems, related to a user account Cosm.com. A free account allows up to 10 sensors Cosm.com updated in real time, and the collected data is saved for 3 months.

Our idea is to connect these sensors to Arduino and Arduino to connect Cosm.com, so that sensors measuring data will defined periods of time and will automatically Arduino sending this Cosm.com to save them. In this way, you can view the current status of sensors and various statistics more (presented in different forms very accessible) from anywhere in the world by accessing its website (or through specific applications for next-generation mobile phones) by the account User related to these sensors.

In addition, this information housed in Cosm.com can also be shared and used by other applications (which has previously been granted access) as data entry procedures. This would allow us, for example, control actuators according to information received from Cosm.com, actuators may be located geographically in a distant place where are the sensors.

To send information to Cosm.com not need any special library Arduino: we only send to server "api.cosm.com" an HTTP request with specific headers. But for these headers are, besides having an account in Cosm.com, we get a couple of data (a "Api Key" and "feedid") that we always include in our sketches accepted for Arduino is authorized to send data to that particular account. To achieve this, the first thing to do is to log on the website of Cosm.com and create a new "device" of Arduino type using the relevant link. In this device we give it a name.

Once created the device, we go to the link "Keys" for the "Api Key" and "feedid". The first is a string that identifies us within Cosm.com and will allow, to specify in our sketches, these are allowed to send data to our own. The second is a number that identifies a particular data flow from our device and allow, to specify in our sketches, these can inform
Cosm.com on which particular sensor is the one actually sending data. For a single device we can define several "feedid", each corresponding to a different sensor, but this is not usual, because in fact, with a single "feedid" you can send data from multiple sensors together, as we shall see in the following examples.

From the central control panel (accessible through the "Console" link) you may modify the characteristics of the newly created device. You can, for example, add geolocation information, add alerts ("triggers") when the values obtained are set out in a certain range, make public the collected data, export data to different formats, add tags ("tags ") to help find that device from all around the world in different categories, etc. You can also observe the public data of other users in https://cosm.com/feeds?status=live Cosm.com.

Imagine now that we have our plate / Arduino Ethernet shield connected to any sensor (light, temperature, etc.) and also through our gateway LAN, Internet. If we send every 10 seconds (for example) the sensor data to a particular account Cosm.com, should write code like the following:

Imagine now that we have our plate / Arduino Ethernet shield connected to any sensor (light, temperature, etc.) and also through our gateway LAN, Internet. If we send every 10 seconds (for example) the sensor data to a particular account Cosm.com, should write code like the following:

```
#include <SPI.h>
#include <Ethernet.h>
apikey const char [] = "The apikey of api.cosm.com was put here";
// Here we have to put our "feed ID"
feedid const char [] = "00000";
// Here we have to put the name of the project const char USERAGENT [] = "My
project";
mac byte [] = {0xDE, 0xAD, 0xBE, 0xEF, 0xFE, 0xED}; IPAddress ip
(192,168,1,177);
EthernetClient MyClient;
/ * Using ip "api.cosm.com" server, the size of our sketch is reduced Flash memory *
/
IPAddress myserver (216,52,233,121);
// Myserver char [] = "api.cosm.com";
// When connected to Cosm.com last time, in milliseconds unsigned long
ultimaConexion = 0;
```

124

```
// Connection status in the last iteration of the loop boolean connected = false;
// Time between connections Cosm.com
const unsigned long interval = 10 * 1000; 10000 milliseconds // void setup () {


if (Ethernet.begin (mac) == 0) {
Serial.println ("DHCP failed Taste fixed IP."); Ethernet.begin (mac, ip);
}
}
void loop () {
int read;
/ * Data to send Cosm.com are strings that always have the same format:
"nombredelsensor, valordelsensor". On behalf of the sensor we can write whatever
we want * /
String message = "sensor1";
// Either I read a sensor, in this case reading analog analogRead = (0);
// Concateno the value of reading the message, to complete mensaje.concat
(reading);
/ * You can add multiple readings from different sensors in the same message,
making each partner nombresensor <-> valorsensor on a different line. Each
appears in a different Cosm.com as "Datastream" within the "feed" used. For
instance:
analogRead otralectura = int (1); mensaje.concat ("\ nsensor2,"); mensaje.concat
(otralectura); * /
```

/ * If you are currently connected to Cosm.com and 10 seconds have passed since the last connection, reconnect and sending the data * /

if (micliente.connected () == false && (millis () - ultimaConexion> interval))

{EnvioDatos (message);

}

}

envioDatos void (String data) {

if (micliente.connect (myserver, 80) == true) {

/ * Send an HTTP request to the web server "api.cosm.com". In this case the request type is GET (not want to see any web page) but PUT type (because we want to send some data to be collected). The other lines of the header must be placed as * /

micliente.print ("PUT / v2 / feeds /");

micliente.print (feedid);

/ * The data to be sent are separated by commas. This is known as CSV ("comma-separated values") format, but may also use other, as JSON or XML * /

micliente.println ("csv HTTP / 1.1."); micliente.println ("Host: api.cosm.com");

micliente.print ("X-ApiKey:");


micliente.println (apikey); micliente.print ("User-Agent:"); micliente.println (USERAGENT); micliente.print ("Content-Length:");

I // sends the number of bytes occupied by the micliente.println message (dato.length ()); micliente.println ("Content-Type: text / csv"); micliente.println ("Connection: close"); micliente.println (); // End of the head of customer

/ * Now it is when really the message is sent with the PUT request data * /

micliente.println (data);

```
delay (100); // I waited a moment and I disconnect micliente.stop ();
} Else {// If I can not connect successfully, I disconnect
micliente.stop ();
}
Anoto // when the connection was made or attempted ultimaConexion = millis ();
}
```

The above code serves as a sample to see how it could implement a standard web client to request the same page every ten seconds (if we change the type of request for a GET PUT request and make minimal changes more). It is left as an exercise.

In fact, using the official Arduino library Cosm.com (https://github.com/cosm/cosm-arduino) would have greatly simplified our code, but then we would not have learned how communication and data transfer with this platform works internally , well aware that we will come to see that many such services Cosm.com (listed in the next paragraph) work in a similar way. In addition to the official bookstore, there are other third-party libraries that perform the same work equally well; In this sense, we can test the "Arduino Cosm Library" available at https://github.com/blawson/PachubeArduino or also the library "PachubeLibrary" available at http://code.google.com/p/pachubelibrary.

Cosm.com is very comprehensive and can perform many functions but not explained here. But it is not the only website of its kind: a very interesting alternative is for example Cosm.com Nimbits (http://www.nimbits.com) platform that offers a library for our official Arduino sketches can interact with it easily, and also offers source code

127

internal so we can mount a server Nimbits our own computer if you wish. Similar services are also Thingspeak (https://www.thingspeak.com) OpenSense (http://open.sen.se) or SensorMonkey (https://sensormonkey.eeng.nuim.ie). In any case, I recommend reading the documentation and guide each of these services to see which one best fits our needs.

Case study: data collection from Cosm.com

As we send our Arduino data from local sensors Cosm.com website, our Arduino can also get data from remote sensors (from our "feeds" or other users that have been released) and use this information as it suits us. In this way, we could control an Arduino installation using a data entry from anywhere in the world. For example, according to the temperature detected in Hong Kong we could light a LED connected to an Arduino located in Montreal.

This data can get them in three different formats: CSV, JSON or XML. Therefore, to retrieve the desired information should be able to understand the internal structure of the chosen format. Fortunately, this is not necessary because what we do is use a library that greatly facilitate things for us. It is downloadable Seller "PachubeLibrary" the http://code.google.com/p/pachubelibrary page. This library allows both very simply sending data from local sensors (which is what we have done in the previous case study), as obtaining remote sensing data (which is what we use it). In both cases provides example code, and shown below is based on one of them. Specifically, the following sketch gets every five seconds the "datastreams" belonging to "feed" and specified "apikeys" and shows the channel number.

```
#include <SPI.h>

#include <Ethernet.h>

#include <ERxPachube.h>

mac byte [] = {0xCC, 0xAC, 0xBE, 0xEF, 0xFE, 0x91};

ip byte [] = {192, 168, 1, 177};

apikey const char [] = "The apikey of api.cosm.com was put here";

// Here we have to put our "feed ID"

feedid const char [] = "00000";

// I think an object that will handle the data received from feedid ERxPachubeDataIn

data (apikey, feedid);

void setup () {Serial.begin (9600); Ethernet.begin (mac, ip);

}

void loop () {

int estadoConexion;

numDatastreams word, i;

// Connect to the feedid

datos.syncPachube estadoConexion = ();

// If the connection is successful, we receive an HTTP 200 (OK) Serial.println
(estadoConexion) code;

// Count how many datastreams has feedid (may be more than one)
datos.countDatastreams numDatastreams = (); Serial.println ("<Sensor>, <value>");

for (i = 0; i <numDatastreams; i ++) {

// I get the name of the sensor datastream "i" Serial.print (datos.getIdByIndex (i));
Serial.print (",");
```

```
// I get the sensor value datastream "i" Serial.print (datos.getValueByIndex (i));
Serial.println ();
}
delay (5000);
}
```

Case study: sending data to Google Spreadsheets

If we want to store the data read by our sensors in an online spreadsheet, so you can perform different operations statistics, graphic display from the information obtained, download this information into our computer or shared with other users (including many other possibilities), one of the most practical ways is to use the "Google Spreadsheets" service, part of the suite "Google Docs". To do this, we need to have a user account Google.

The trick is to create a "Google Form" to receive different data sent by our Arduino. "Google Form" is a service that allows you to generate online forms very quickly and easily, ideal for surveys or questionnaires. The good thing about "Google Form" is automatically imported into "Google Spreadsheets" the received data, without having to do anything.

To create a form of Google, we turn to http://docs.google.com and click on the "Create-> Form" button. We must give a title to the form and all questions we believe. The questions that we add (using the "Add item" button) must always be of type TEXT. The titles of the questions will be the names of the columns in the spreadsheet, and each response obtained is stored in a separate row of the spreadsheet. Once completed the form, we click on the "Save" button. At the bottom of the page will appear us a link to display the form and finish. We click on it to look at the address bar of the browser, because there we will see the "formkey" form, which is a long string of numbers and letters that identifies the form of all others. This formkey the sketch we include in our Arduino to our board to properly interact with it.

We must also set another important detail for our sketch: TEXT type boxes in the form of Google usually have an identifying internal name as "entry.0.single" (for the first question), "entry.1.single" (for the second), etc. However, if we modify the basic structure of these names may change form; to ensure they have not done so, it is good idea to check the HTML code of the form and finished and presentable. If we use the free Firefox browser, cross-platform, this can be done simply by simultaneously pressing the CTRL and U.

We have created the form, and therefore it is able to store in "Google Spreadsheets" data entered interactively. But if we want our Arduino who enter them independently, we know that this introduction can be done by sending Google the address: https: //spreadsheets.google.com/formResponse formkey = MIFORMKEY & IFQ
& NOMBREINTERNOPREGUNTA1 = VALUE & NOMBREINTERNOPREGUNTA2 = VALUE & sub = mit Submit

131

The following code shows how to save about a Google Spreadsheet every two seconds the values of two analog sensors connected to the No. 0 and No. 1 pin If there were no sensor connected to those pins, random values are obtained from ambient noise .

```
#include <SPI.h>
#include <Ethernet.h>
// It is to substitute the "formkey" Google Form our formkey char [] =
"DdMBUd3xmTQ52Yvx2XZ01V83VUp2U06EQM";


mac byte [] = {0x90,0xA2,0xDA, 0x00,0x55,0x8D};
ip byte [] = {} 192,168,1,177;
subnet byte [] = {} 255,255,255,0; servdns byte [] = {8,8,8,8}; gateway byte [] = {}
192,168,1,1;
myserver char [] = "spreadsheets.google.com"; Data String = "";
EthernetClient MyClient;
void setup () {
Ethernet.begin (mac, ip, servdns, gateway, subnet);
// I hope to give you time to initialize the Ethernet shield delay (1000);
}
void loop () {
// I send chain concatenating different parts data = "entry.0.single =";
data = data + analogRead (0); data = data + "& entry.1.single ="; data = data +
analogRead (1); data = data + "& submit = Submit";
if (micliente.connect (myserver, 80)! = 0) {micliente.print ("POST / formResponse
formkey =?"); micliente.print (formkey); micliente.println ("& IFQ HTTP / 1.1");
```

```
micliente.println ("Host: spreadsheets.google.com");

micliente.println ("Content-Type: application / x-www-form-urlencoded");

micliente.println    ("Connection:    close");    micliente.print    ("Content-Length:");

micliente.println (datos.length ()); micliente.println (); micliente.print (data);

micliente.println ();

}

delay (1000);

if (micliente.connected () == true) {micliente.stop (); }

delay (1000);

}
```

In the previous sketch, for sending data to Google, we have made use of the POST method instead of GET method seen in previous examples. The difference between the two methods is that GET data received by the form is sent as a part of the address of the requested page (and thus, are visible on the address bar of your browser) and data is sent POST

as a header over client at the end of all other and located between blank lines.


Case study: sending notifications to Pushingbox.com

Pushingbox.com (http://www.pushingbox.com) is an online service notifications. So that it can prove need previously to have a user account and a Google Arduino prepared plate having sensors to detect an event. The board shall execute a specific sketch, which will send a notification to Pushingbox.com whenever that event occurs. The interesting thing is that Pushingbox.com handles forward this message to the (multiple) targets that we set: can issue a notification to a Twitter account, a mailbox, or a specific application for mobile phones last generation, among others. Thus, we can be informed anytime, anywhere of what happens in our installation Arduino.

The steps for using this service are as follows. First we log in with a user of Google and go to the section "My services". There we click on "Add a service". We will see that we have several possibilities of "receivers" of the notifications sent by Pushingbox: we can choose a specific mailbox, a Twitter account or different applications for next generation mobile (specifically, Prowl Pushme.to phone or iPhone, or Notifry "Notify my Android" Toasty Android or Windows Phone mobile phones). For every "recipient" selected (can choose several at once) us a "Submit" button and we click we will lead to a different process depending on the type of "receptor" chosen appear. In the case of using mobile applications in this way we enter the "API Key" concrete that was generated to install the application on your device.

After this step, we must continue to go to "My scenarios" section. Here we give a name to each scenario we want to create and clicaremos in "Create a scenario". Creating a scenario allows us to basically get a "DeviceID" is an identifying string that we introduce in the sketch executed by Arduino, which will allow us to identify in Pushingbox. Once created the stage (or stages) desired, we will have to decide which "receiver" as defined in the previous step we use. To do this, we click on "Add an action". Depending on the "receptor" we ask various details (if an email tell us what matters and what text we want to add, if Twitter is sending to tell us what message we want to add, etc.). And that's it.

We have the "Test" button to test the sending of notifications for that action to be successful, but the interesting thing is that this shipment is performed by the Arduino.

To do this, we can use the sketch example shown below. To prove we have a button connected to digital input pin No. 3 of our Arduino. Each time you press, the sketch notify PushingBox (and this renotificará what the "receptors" that we have set for the stage associated with the specified DevID in the code). Every time you stop pressing a notification will also be held. Obviously, this very basic example is a reference only to add here from various sophisticated sensors.

#include <SPI.h>

#include <Ethernet.h>

mac byte [] = {0x00, 0xAA, 0xBB, 0xCC, 0xDE, 0x19};

/ * The following line is to be put for the stage to be monitored DevID. You can use multiple DevIDs in one sketch (associated with different pins on the Arduino board).

* /

```
DEVID1 String = "Tu_DevID_Aqui";

myserver char [] = "api.pushingbox.com"; estadopulsador boolean = false;
EthernetClient MyClient;

void setup () {

// Pin digital input where I have connected the button pinMode (3, INPUT);

/ * If there is error, I do not do anything and if everything is ok, continue. I assume
that there is a DHCP server on my LAN that gives me all data connectivity * /

if (Ethernet.begin (mac) == 0) {while (true) {;}}

delay (1000);

}

void loop () {

// If I pressed the button and it was not until tight

if (digitalRead (3) == HIGH && estadopulsador == false) {

estadopulsador = true;

// Sending the notification to Pushingbox enviarAPushingBox (DEVID1);

}

// If I stop press the button

if (digitalRead (3) == LOW estadopulsador && == true) {

estadopulsador = false;

// Sending the notification to Pushingbox

enviarAPushingBox (DEVID1);

}

}

// Function notifications to Pushingbox void enviarAPushingBox (String devid) {

if (micliente.connect (myserver, 80)) {micliente.print ("GET / pushingbox devid =?");
```

```
micliente.print (devid);

micliente.println ("HTTP / 1.1"); micliente.print ("Host"); micliente.println (myserver);

micliente.println ("User-Agent: Arduino");

micliente.println ();

}

micliente.stop ();

}
```

Arduino Shields alternative to Ethernet

Besides the official shield, other third shields also add Ethernet connectivity (TCP / IP) to an Arduino UNO and we can choose according to their characteristics. Shields Wiz5100 incorporating the same chip as the official shield are for example the "Ethernet shield" of DFRobot or "Ethernet shield" of Seeedstudio (which, however, do not include microSD card socket).

Other shields incorporate a module already integrated PoE, as the "Ethernet shield with POE" of Freetronics. This shield has a particular pin (suitably marked) which allow to adapt their behavior depending on the value of the input voltage carried by the PoE signal: if the voltage is up
12 V, only they will have to place a couple of bridges plastic ("jumpers") on

these pins and you're done (with this we will make the pin PoE "+" the shield to pin "Vin" Arduino and pin connects "-" connects to the pin "GND"); if the voltage is up to 28 V, it is necessary then be coupled to pins of the shield proper voltage regulator such as "PR28V" (also manufactured by Freetronics own) to avoid burning the plate. This regulator provides up to 1A of current, an output voltage of 5 V to 7 V and can receive an input voltage up to 28 V, which is not compatible with the 802.3af standard (which allows voltages up to 48 V) and therefore is not prepared to be connected to PoE switches business. To connect this shield to such switches (and then receive power through the PoE signal voltage up to 48 V), it is necessary to use other

more capable controller (such as "Regulator 802.3af PoE" also provided by the Freetronics own).


Another shield similar to the previous example is the "PoEthernet shield" SparkFun. Has a pair of holes marked "GND / -" and "Vin / +" welding pin in which there will place a couple of "jumpers" (as in the shield above). These pins allow properly receive the signal by PoE Ethernet cable, signal that can be regulated to 5 V and 3.3 V can not be applied, however, PoE signal with a higher voltage of 12 V. This shield also includes microSD socket accessible through official "SD" library.


If you do not want to use a full shield, but want to use an independent external Wiz5100 module (with built-in RJ-45 connector), we can obtain the "WIZnet W5100 Ethernet / Network Module 'from IteadStudio. This module communicates with the outside via the SPI protocol, so it should be programmed by the official bookstore "SPI" Arduino. Another plate similar to previous Sparkfun is distributed with the product code 9473.

138

On the other hand, there are shields that do not include Ethernet Wiz5100 chip, but a different chip, Microchip ENC28J60 (also connected to the plate by SPI). This chip allows a much more detailed analysis of network communication control and greater freedom in the use of different protocols, Wiz5100 freedom that is not able to offer. However, it requires a somewhat higher knowledge to be programmed. Technically, this is because the Wiz5100 chip features a stack of TCP / IP protocols already built into the hardware, so that the user does not have to program from scratch; this offers the advantage of not having to "hands dirty" working at a low level, but provides the disadvantage of not being as flexible and versatile as the ENC28J60 chip.

Examples of shields of incorporating the ENC28J60 chip are the "Gate 0.5" Snootlab or "IE shield" of Iteadstudio (which also incorporates a PoE module and a microSD socket), which have been programmed with a different library of Official bookstore "Ethernet" Arduino. The library used is called EtherCard (http://jeelabs.net/projects/cafe/wiki/EtherCard. Another less known is available in https://github.com/turicas/Ethernet_ENC28J60 library. You can also add the ENC28J60 chip functionality to our Arduino UNO by SPI module (as the "ENC28J60 Ethernet module" of Iteadstudio), without therefore using a full shield.

Communication network using a standard Arduino UNO

You can connect a Arduino UNO to an Ethernet network without using the shield and Arduino Ethernet or Ethernet card itself: only with the help of a computer is already possible. Specifically, our plate connecting one via USB to a computer with access to the network, we can use this as an intermediary to forward messages to each outer plate (and also counter-clockwise).

139

To do this, we need to run a program on the computer to set the USB port serial communication with the Arduino (to receive data from it-from sensors, normalmente- and also to send signals -a Control actuators normalmente-) to Once you set the Ethernet port on the network communication with the rest of the LAN and the Internet. That is, acting between the network and an Arduino with no capacity as a "bridge" to connect to itself. These programs are generically called "proxies-network series".

A "proxy-network series" normally we develop ourselves, since its functionality will depend much of the design of our project and the type of message we want to transmit or receive from abroad. Therefore, we must know a programming language that allows us to create such an intermediary program that simultaneously contacting Arduino via serial port and the rest of the world via Ethernet.

It completely escapes the scope of this book deepen the use of other languages different programming Arduino itself, but as a reference, we can appoint languages that have been sufficiently tested in these circumstances. One is naturally Processing, Arduino language much like focusing on the development of visual and animated applications. It incorporates a library called "Serial" (http://processing.org/reference/libraries/serial) and another library called "Network" (http://processing.org/reference/libraries/net) for linking a same program both ends of the communication (serial and Ethernet).

Processing a different language, but is also free and cross-platform Python (http://www.python.org). This language is multipurpose and can be used to develop binary applications ("executable") or web type. In the present case, for communication via Ethernet there are several possibilities, but the most versatile and powerful is to use the "Socket", integrated into the default language interpreter. For communication via serial

change will not have any default library, so we should download and install a third party. The most recommended is the "PySerial" library available in http://pyserial.sourceforge.net.

Another language, PHP is free and multiplatform (http://www.php.net), multipurpose language that can also be used to develop binary applications or (especially) Web type. It provides a syntax much like the Arduino language Python, because, like him, is based in C. In the present case, for communication via Ethernet there are several possibilities, but the most versatile and powerful is to use the "Sockets", integrated into the default language interpreter. For communication via serial however, we do not have (as in Python) in any library by default, so you should download and install a third party. The best practice is available Bookseller http://code.google.com/p/php-serial "PhpSerial".

On the other hand, very common when the intermediary computer settings are describing is to collect data from Arduino to store them on a server in the remote database is used. In that case, rather than a generic Ethernet library, the programming language that is developed with the series-network proxy must provide specific connection to work with specific servers and database libraries. All the above languages (Processing, Python or PHP) feature this functionality.

If you do not have knowledge (or time) sufficient to develop our own proxy-network series generic, we can also choose one already developed and ready to be installed and used at the time. So we will not have to worry about developing anything "by hand" and we only deal with learning to use that particular software; the trouble is that such software may not suit our particular needs. The idea is in all cases the same: the sketches of our Arduino communicate with our computer using the Serial object and our computer, by proxy we have installed, will be responsible to open a TCP / IP (configurable) port to do available to the world.

An example of series-network proxy (functional only on Windows systems, however) is Bloom, https://sensormonkey.eeng.nuim.ie available. A serial-network proxy running on Linux systems is Ser2net (http://ser2net.sourceforge.net). A series-network proxy running on both systems is SerProxy, available http://www.lspace.nildram.co.uk/freeware.html.

# COMMUNICATION THROUGH WI-FI

## What is Wi-Fi?

The apparent operation of a Wi-Fi network is very similar to an Ethernet network, only without cables. However, in addition to IP addresses and MAC addresses, in this wireless technology it has to take into account other concepts:

IEEE802.11 standard "Wi-Fi" is based on this standard, which is actually a set of standards. Depending on compatibility with one or more of these standards, we will find devices that can be part of networks Wi-Fi 802.11b, 802.11g or 802.11n among others. The most important difference between them is the speed of data transmission (up to 11

Mbit / s, 54 Mbit / s and 600 Mbit / s, respectively), but the three

specifications operating in the 2.4 GHz band (in addition 802.11n 5.4 GHz) and signal range reaches the hundred meters.

Access Point (AP): A hotspot is a wireless network computer (either a computer with the right software, or a specific hardware device) that is responsible for centrally manage all communications devices that form the Wi-Fi network. Not only used to control internal communications network, but also a bridge for communication with external networks (Ethernet and Internet networks), as a "signal transformer" between wireless and wired networks. Wi-Fi networks that do not have access point may exist; in this case, communications are carried out directly between the different terminals forming the network and its structure is often called "ad-hoc" (or also as equals "-" peer-to-peer "- or IBSS ). Wi-Fi networks that offer access point have a structure called "infrastructure" (or "BSS").

143

Mode: A Wi-Fi device may have a particular role within the network, and this is configured by setting its mode of operation. The Station or so ("Managed") is the way in which a device is merely a client connecting to an access point for connectivity. AP mode or ("Master") is the way in which a device can work himself as an access point (if you have the proper firmware). There are more exotic modes we will not see.

SSID: is data issued by the access point that identifies the wireless network to which it belongs. In other words, is the "network name" terminals are able to see in order to connect. The SSID can be changed by accessing the access point settings. Do not confuse with the BSSID, which represents the MAC address of the access point.


Channel: the band of electromagnetic frequencies in working a Wi-Fi network (the 2.4 GHz band, generally) is divided into multiple channels. Specifically, the standard divides the 2.4 GHz range in 14 separated by 5MHz channels (although each country has its own restrictions on the number of channels available: in Europe can only be used from channel 1 to 13, for example ). The problem with this arrangement is that each channel need 22 MHz bandwidth to operate, and as seen in the figure, this produces an overlap of several adjacent channels (e.g., channel 1 overlaps with channels 2 , 3,

4 and 5). The consequence of that the devices emit in overlapping ranges is that they generate interference each other, hampering connectivity and network speed. So, as far as possible, you should choose different channels to avoid this (the most common is to choose channels 1, 6 and 11).

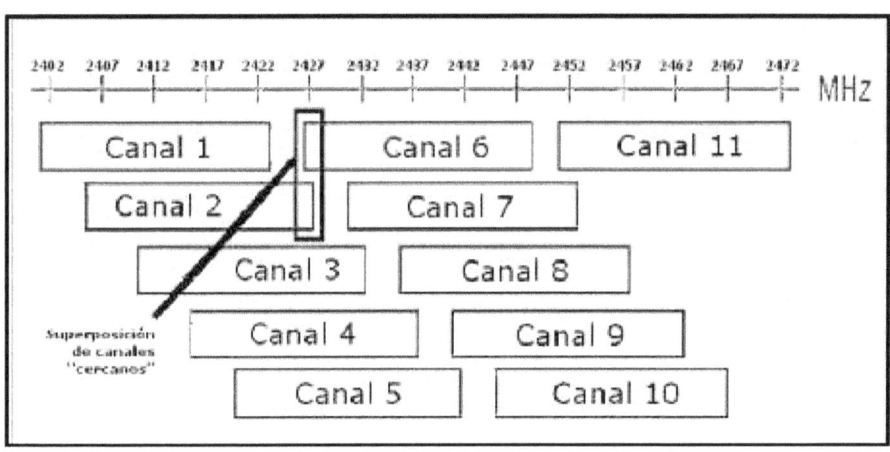

Encryption Algorithm: communication between devices forming a Wi-Fi network can be encrypted, so that data transmitted between them can not be known devices outsiders. So it provides security and confidentiality some communications that are easily captured by itself (since the medium is air). There are different encryption methods: the (safer than WEP but not enough) WEP (not recommended due to its weakness against attacks decryption), the system or WPA-TKIP the latest WPA2-CCMP (aka WPA2-AES and 802.11i implements enhanced data protection and improved access control), among others. Within the method WPA2-CCMP can choose various storage technologies passwords: the PSK (also called "WPA2 Personal") or EAP (also called "WPA2 Enterprise"); Personal WPA2 is more suitable for home wireless facilities due to rapid and simple configuration, and WPA2-Enterprise is designed for corporate installations where the extra security offered by a central server passwords (RADIUS type) is required. In any case, all encryption methods studied in this book require the introduction into the access point setting a password (or "pass key"), which should be known by the devices that want to join him.

Using Arduino WiFi Shield and the official bookstore WiFi

If to use Wi-Fi communication we chose to work with official Arduino shield (the "Arduino WiFi Shield"), we set our sketches with the help of the "WiFi" library. This library is quite similar to the "Ethernet" library view in previous sections, so only the main differences will be mentioned and referred to the study of the "Ethernet" library for details.

Once imported the library, the first thing we do is initialize by WiFi.begin () function, which is somewhat equivalent to the already known Ethernet.begin () measure. This function can be written in three different ways, returning all the predefined constant value WL_CONNECTED if you can connect to the Wi-Fi network or WL_IDLE_STATUS specified value or not.

WiFi.begin (SSID): where "SSID" is the SSID of the Wi-Fi network you want to connect. This way of writing WiFi.begin () is what we use when the Wi-Fi network is open type (unencrypted).

WiFi.begin (ssid, pass): where "pass" represents the password for the Wi-Fi network protected with WPA2-Personal to which you want to connect.

WiFi.begin (ssid, index, key): if you want to connect to a Wi-Fi network protected by WEP, we must provide two pieces of information: "index" and "key". The WEP passwords are strings of 10 or 26 hexadecimal characters ("key" parameter) and as may exist in four different
same network, the "index" parameter is used to indicate which one (0, 1, 2 or
3) you want to use.

As you can see, in WiFi.begin () can not specify any fixed IP (or mask, or gateway and dns servers) because it is assumed that all this information is dynamically assigned by DHCP access point to which connect the shield at that time.

Other popular WiFi general library functions are:

WiFi.disconnect (): disconnect the shield of the current network. No parameters or return value.

WiFi.scanNetworks () scans the Wi-Fi networks and returns the total number of found networks -dato of byte-type. No parameters.

WiFi.SSID () return value is a string that contains the SSID of the network you are currently connected to the shield. Optionally, you can specify a parameter byte-type -of representing the numeric position (0,1,2 ...) in the list of found networks by WiFi.scanNetworks () that has the network which you want to know your SSID (although not the network you're currently connected).

WiFi.BSSID (): no return value, but a mandatory parameter must be of type array of byte type 6 positions. There the MAC address of the access point to which the shield is connected at this time is saved. The element 0 of the array corresponds to the MAC byte more to the right, and 5 more to the left.

WiFi.RSSI () returns a value type long- -of representing the strength of the signal received in the current connection (called "Received Signal Strength", measured in dBm). Optionally, you can specify a parameter byte-type -of representing the numerical position (0, 1, 2 ...) in the list of found networks by WiFi.scanNetworks () that has the network that want to know this information (although not the network you're currently connected).

WiFi.encryptionType (): returns a byte-type -of value representing the type of encryption to the existing network. 2 equals WPA-TKIP, WPA2 4 to
CCMP, WEP and 5 to 7 to open network. Optionally, you can specify a parameter byte-type -of representing the numerical position (0, 1, 2 ...) in the list of found networks by WiFi.scanNetworks () that has the network that want to know this information (although not the network you're currently connected).

WiFi.macAdress (): no return value, but a mandatory parameter must be of type array of byte type 6 positions. There's MAC address will be saved shield itself. The element 0 of the array corresponds to the MAC byte more to the right, and 5 more to the left.

WiFi.localIP (): returns the current IP address of the shield. This return value must be previously declared type "IPAddress" (which no longer a simple array 4 numerical positions). This function has no parameters. It would be equivalent to Ethernet.localIP () function.

WiFi.subnetMask (): Returns the current network mask of the shield. This return value must be previously declared type "IPAddress". This function has no parameters.

WiFi.gatewayIP () returns the IP address of the gateway currently configured on the shield. This return value must be previously declared type "IPAddress". This function has no parameters.

As with the Ethernet library in the library WiFi objects have both server and client-type objects, and both libraries functioning and behavior is exactly the same:
To create a server object: we must declare by a line like this: myserver WiFiServer (80) ;, where "myserver" is the name we have chosen for the purpose of WiFiServer type, and the number in parentheses represents the port to open waiting for incoming client connections. From here, we can use the functions already known miservidor.begin (), miservidor.available () -which, in this case, it returns an object of type "WiFiClient" - miservidor.write (), miservidor.print ( ) and miservidor.println ().

To create a client object: we must declare by a line like this: WiFiClient MyClient; where "MyClient" is the name

we have chosen for the object type WiFiClient. From here, we can use the functions already known micliente.connected (), micliente.connect (), where you can specify the server by its IP or DNS-name, micliente.write (), micliente.print () , micliente.println (), micliente.available (), micliente.read (), micliente.flush () and micliente.stop ().

Then we show a few sketches that affect the novelty of this library regarding the Ethernet library. The examples presented above in the corresponding section Ethernet library to remain perfectly valid library with WiFi: only has to change the way of establishing a connection with WiFi.begin () and the type of objects used (WiFiServer and WiFiClient).

8.4 Example: The following code does not connect to any Wi-Fi network: it does is show the channel number detected networks (and gift, shield the MAC address). We must clarify that possibly not detect the shield as many networks as a computer, as these often incorporate a larger antenna.

```
#include <SPI.h>
#include <WiFi.h>
void setup () {Serial.begin (9600);
printMacAddress (); // Shows the MAC address of the shield
}
```

150

```
void loop () {
listNetworks (); // Displays information available WiFi network delay (10000);
}
printMacAddress void () {byte mac [6]; WiFi.macAddress (mac); Serial.print ("MAC");
Serial.print (mac [5], HEX); Serial.print (":"); Serial.print (mac [4], HEX); Serial.print
(":");  Serial.print  (mac  [3],  HEX);  Serial.print  (":");  Serial.print  (mac  [2]  HEX);
Serial.print (":"); Serial.print (mac [1], HEX); Serial.print (":");
Serial.println (mac [0], HEX);
}
listNetworks void () {
WiFi.scanNetworks numSsid byte = (); Serial.print ("Number of available networks");

Serial.println (numSsid);
/ * Sampling rate found each network, besides its SSID, signal strength and
encryption used * /
for (int i = 0; i <numSsid; i ++) {
Serial.print (i); Serial.print (")"); Serial.print (WiFi.SSID (i)); Serial.print ("\ tSeñal (in
dBm):"); Serial.print (WiFi.RSSI (i)); Serial.print ("\ tCifrado:"); Serial.println
(WiFi.encryptionType (i));
}
}
```

8.5 Example: The following code shows how to connect the shield to an open WiFi network, called "mired".

```
#include <SPI.h>
#include <WiFi.h>
char ssid[] = "mired";
int status = WL_IDLE_STATUS;
void        setup()        {
   Serial.begin(9600);
   status = WiFi.begin(ssid);
   if    (status    !=    WL_CONNECTED)    {
      Serial.println("Error en la conexión");
   } else {
      Serial.println("Conectado");
   }
}
void loop() {}
```

```
#include <SPI.h>
#include <WiFi.h>
char ssid[] = "mired";
char pass[] = "12345678";
int status = WL_IDLE_STATUS;
void        setup()        {
   Serial.begin(9600);
   status = WiFi.begin(ssid, pass);
   if    (status    !=    WL_CONNECTED)    {
      Serial.println("Error en la conexión ");
   } else {
```

```
      Serial.println("Conectado");
  }
}
void loop() {}
```

8.7 Example: The following code shows how to connect the shield to a WiFi network name "mired" and encrypted using WEP with a key of 10 hexadecimal characters and index 0.

```
#include <SPI.h>
#include <WiFi.h>
char ssid[] = "mired";
char key[] = "ABBADEAF01";  //La clave de 10 caracteres
int keyIndex = 0;        //El índice de la clave dentro del AP
int status = WL_IDLE_STATUS;
void        setup()        {
  Serial.begin(9600);
  status = WiFi.begin(ssid, keyIndex, key);
  if     (status     !=     WL_CONNECTED)     {
    Serial.println("Error en la conexión ");
  } else {
    Serial.println("Conectado");
  }
}
void loop() {}
```

a set of network parameters (ip, mask, gateway, etc.) thanks to the access point to which it connects. In our sketches we can use the following function to know at all times the value of these parameters.

```
printWifiData void () {
WiFi.localIP IPAddress ip = (); Serial.print ("IP Address"); Serial.println (ip);
IPAddress WiFi.subnetMask subnet = (); Serial.print ("Network mask"); Serial.println (subnet);
IPAddress WiFi.gatewayIP gateway = (); Serial.print ("Gateway");
Serial.println (gateway);
}
```

Example 8.9: Once connected we also know some facts Example 8.8: Once connected to a WiFi network (of any kind), the shield will acquire the network that we have built (ie the SSID of the network, strength signal to the access point, the encryption algorithm used and the MAC of the access point) by executing on our sketch of the following function:

```
printCurrentNet void () {Serial.print ("SSID"); Serial.println (WiFi.SSID ()); Serial.print ("Signal Strength (RSSI):"); WiFi.RSSI long rssi = (); Serial.println (RSSI);
Serial.print ("Encryption type");
WiFi.encryptionType encryption byte = (); Serial.println (encryption, HEX);
I show the MAC // access point to which I am connected bssid byte [6];
WiFi.BSSID (BSSID); Serial.print ("BSSID"); Serial.print (bssid [5], HEX); Serial.print (":"); Serial.print (bssid [4], HEX); Serial.print (":"); Serial.print (bssid [3], HEX);
Serial.print (":"); Serial.print (bssid [2] HEX); Serial.print (":"); Serial.print (bssid [1], HEX); Serial.print (":");
Serial.println (bssid [0], HEX);
}
```

Other shields and modules that add Wi-Fi connectivity

Besides the official shield, other shields also add Wi-Fi connectivity to Arduino. In this category we find for example the "Wifly Shield" Sparkfun (product # 9954), that can connect to Wi-Fi networks 802.11b / g using various encryption systems thanks to Wi-Fi RN-131C module that incorporates (Manufacturer Roving Network). Like the official shield, it communicates with the motherboard through the SPI pins, but unlike that, the "Wifly Shield" operates at 3.3 V regulated. Another detail of this shield is that it has a small area of prototyping. The easiest (although not the only) way to configure and manage this shield is using available https://github.com/sparkfun/WiFly-Shield bookstore.

Another interesting shield is the "Wifi shield" of DFRobot, which incorporates the Wi-Fi module WizFi210 Wiznet. This module provides 802.11b / g / n with the possibility of also using various encryption systems. Also, this shield allows coupling thanks to an external antenna connector type "U.FL". The configuration of this shield is somewhat different from the previous ones: to establish the connection data of the shield (SSID to associate with the possible password, request indication whether or not IP via DHCP, etc.) is to be used type program "serial terminal" and send a series of specific commands. Once this is done, the shield is already configured by default to connect to a predetermined network. From there we can already send and receive data through it using our sketch simply Serial.print () and Serial.read (). Windows users can use the graphics program WizSmartScript (downloadable from the official website of Wiznet) it helps to have this shield configured in a few steps without having to know any specific command. For details on setting up and using this shield, I refer you to the information available on the website of the product.

Another example is the shield "wireless Arduino Shield", Open-Electronics, which incorporates an integrated antenna and chip Microchip ZG2100 (all encapsulated as a module, the MRF24WB0MA). This shield allows use modes 802.11b / g / n, also it includes a microSD socket, operates at 3.3 V, and communicates with the board via SPI. In order to configure and use you can use a specific library available http://code.google.com/p/wifi-shield-oe.

Another shield that incorporates the same wi-fi module Microchip is "CuHead Wifi Shield" from Linksprite, which also uses the SPI protocol to communicate with the board, but must be programmed with an own library, downloadable from https: / /github.com/linksprite/ZG2100BasedWiFiShield. Linksprite also builds and sells other wi-fi shields, like the "Anaconda" (which can be controlled simply by Serial.print () and Serial.read ()) or so-called "Juniper" (with wi-fi module GS1011 GainSpan manufacturer). In any case, I refer to the consultation of the respective web pages to know the technical details of configuration and use of each of these shields.

A similar shield is the "Arduino WiFi Shield" from Olimex, which incorporates the base to include a Wiznet Wiz610wi module (the module itself and the antenna has to be purchased separately). This module is able to use the 802.11b / g modes can be programmed with a specific library, downloadable from the website of Olimex. There you can also access your guide to installation, configuration and use, as well as download several code examples.

Another shield that allows connectivity 802.11b (and several encryption methods) is called "Hydrogen" DIY Sandbox. It contains a microSD socket and can be programmed using the library "Wirefree" available at the https://github.com/diysandbox/Wirefree page. The same manufacturer also offers a board called "Platinum" which is compatible with the Arduino UNO and has the ability to connect via Wi-Fi without any shield (in fact, offers the same features as

156

the shield "Hydrogen" - same encryption methods, same microSD socket, same programming library, etc. but in the form of a single plate). This board is programmed via FTDI.

Another shield is "REDFLY" of Watterott. This shield can connect to networks 802.11b / g / n / i type with the same encryption methods as other shields already mentioned. No microSD socket. It is programmed using the library available in https://github.com/watterott/RedFly-Shield.

If we talk about modules, Sparkfun marketed under the product code
10050 a module called "Wifly GSX Breakout" which includes the same chip that comes
in his "Wifly Shield", the RN-131C. This breakout insert in its most basic configuration only requires four connections: power (3.3 V), land, RX (TX by a pin of the Arduino software, to not interfere with the serial communications hardware) and TX ( the Arduino pin RX), and you can be programmed using the library "Serial Wifly" available at http://sourceforge.net/projects/arduinowifly.

Other modules Wi-Fi (cheaper than before) are called RN-XV, also distributed Sparkfun with product codes 11047, 11048 and 10822. These products differ in the type of connector provided for the antenna: the first provides a connector type "SMA" to plug a compatible antenna (such as might be # 145 or 558 number product), the second provides a connector type "U.FL" to plug a compatible antenna (as might for example be the product number 11320) and the third straight antenna type provides a "wire". Bridging these differences, all these breakout plates allow networks to connect to 802.11b / g, in its most basic configuration, require the same four connections: power (3.3
V), land, RX (TX by a pin of the Arduino software, to not interfere with the
series hardware) and communication TX (the Arduino pin RX). All can be programmed using the same library used with the "Wifly Shield" Sparkfun (although incorporating another chip, the RN-171).

157

The advantage is RN-XV modules having the same shape and pin arrangement that XBee, so, in fact, could be used coupled to the official "Arduino Wireless Shield". This feature makes them ideal when we have a facility already assembled XBee modules and desire to make a change to the Wi-Fi technology as quickly and easily as possible: simply replacing a module on the other (and modifying the relevant sketches, clear ) and should be sufficient.

## Communication via Bluetooth

## What is Bluetooth?

Bluetooth (http://www.bluetooth.org) is the name of an industrial specification (officially standardized under the name IEEE 802.15.1) that defines the characteristics of a type of short-range wireless networks. Its main use is to provide a communication protocol between different consumer electronic devices (computers, printers, telephones, digital cameras, audio devices, etc.) relatively close (a few meters away) without having the need to carry a explicit control by the user of network addresses, permissions and other typical aspects of traditional networks. The main advantage of using Bluetooth is to simplify the discovery and automatic configuration of nearby devices, as these can be shown each other their services independently.

The Bluetooth standard used to transmit voice and data link radio frequency in the ISM 2.4 GHz band. The bands ISM ("Industrial, Scientific and Medical") are reserved internationally for non-commercial use of radio frequency electromagnetic. This means you can be openly used worldwide without a license, simply respecting the regulations that limit the amount of transmitted power. There are other wireless technologies using the same ISM band, such as "Wi-Fi" wireless standard (based on the IEEE 802.11 specification, http://www.ieee802.org/11) also uses precisely band 2.4

GHz, as we know. However, Bluetooth and Wi-Fi cover needs

different: Wi-Fi uses a higher output power signal, leading to

stronger, faster, safer and more far-reaching, but requires a certain previous configuration (similar to a traditional Ethernet network) connections.

Within the Bluetooth standard they are defined so-called "device profiles". Each "profile" is superimposed an additional protocol to the core Bluetooth protocol that determines how the device will communicate that to implement it. That is, depending on the profile used, a Bluetooth device will be able to receive / transmit a particular data format and not in another. This means that for two Bluetooth devices to communicate with each other have to use the same profile is not enough to "be alone Bluetooth".

A device may implement one or more profiles, and therefore act in one way or another depending on the use to be given. The most common profiles within the ecosystem Arduino are two: the SPP (Serial Port Profile) and HID (Human Interface Device Profile). The first serves for the device that implements it can establish communication with other devices such series (the style of the UART chips); this is the profile with the modules and Bluetooth shields studied in this section. The second is for the device that implements it can behave as a mouse, keyboard, joystick or a button (and other devices of low latency and consumption). In fact, the Bluetooth HID profile is very similar to HID protocol defined in the USB standard (which implement the Leonardo and Due plates by libraries "Mouse" and "Keyboard").

159

Bluetooth connectivity modules added

If we give our Arduino UNO ability to communicate via Bluetooth, we connect a receiver / transmitter Bluetooth module. These modules typically work as slave devices, so should be the other end of the communication (typically a computer or a mobile phone with Bluetooth capability) which operates as a master device. That is, it will always be the computer / mobile who initiates the connection to the Arduino and not vice versa. If our computer should not have any transmitter / receiver integrated Bluetooth, we could couple it one external, such as Product No. 9434 Sparkfun.

Sparkfun distributes various modules, called generically "BlueSMiRF". Two of them (Product No. 158 and No. 10268) incorporating the RN-41 chip, which allows connections up to distances of 100 meters. The difference between these two modules is that the former includes a connector type "SMA" to plug a compatible antenna and the second antenna includes already integrated into the module PCB.

Other modules "BlueSMiRF" products are the No. 10269 and No. 10393. Both incorporate the RN-42 chip and antenna are integrated into the PCB and module. The main difference is that the second module is specifically designed for use with the Arduino Lilypad Arduino Pro or by FTDI connection. On the other hand, the main difference between the RN-42 and RN-chip 41 of the above modules is that the RN-42 is a "Class 2" while the RN-41 is a "Class 1". A device is "Class 2" means that the maximum useful sensing distance and data transmission is only about 10 meters, but in return, its power consumption is much lower.

In any case, all these modules are configured and used similarly. Basically what you do is transform received from the outside into a serial signal for delivery to the ATmega328P RX Arduino Bluetooth signal, and transform the TX signal generated by the signal ATmega328P a Bluetooth ready to send abroad. Therefore, once it established the relevant wiring, handle data transfer via Bluetooth is as simple as using functions Serial object (specifically, Serial.print () to send abroad and Serial.read () to receive it) . The speed of communication between the modules ATmega328P and BlueSMiRF default is 115200 bits / s, so this value must be specified in Serial.begin ().

The BlueSMiRF modules have 6 connectors. Two of them will not use at all: it is labeled as "CTS" and "RTS". The other connectors are trivial: the power connector "VCC" must be plugged to a power of 3.3 V (5 V but accepts them as well), the "GND" grounded connector "TX" connector must be connected to pin female "RX" Arduino and the "RX" connector must be plugged into pin-female "TX" plate. In addition, there may be something distanced seventh connector of the above labeled "PI04" used to reset the factory settings of the module (this also will use it). However, to avoid problems when configuring these modules, wiring should not "hit" the four connectors (VCC, GND, TX and RX), but must follow these steps:

1. Load on the Arduino board in the usual way (via USB) code we want to run.

2. Feed plugging the module only the "VCC" and "GND" connector and place it near a Bluetooth master device (a computer or a mobile phone) with which you want to communicate. You can do this with more than one master device if desired.

3. The module should be detected by the master device. Bluetooth management software that is installed on this (all operating systems include one "factory") then request a password to communicate with the module. The default password of all modules sold by Sparkun BlueSMiRF is "1234". After entering the key, the master device will automatically recognize the module and a serial port over whose specific

name depends on the operating system of the master device (eg, Windows Bluetooth modules are detected as "COMx" ports on Linux devices / dev / rfcommX, etc). Fortunately, this data will generally not be necessary to know.

4. Connect the module to the Arduino (ie, the RX and TX lines). From here we already have an autonomous Arduino able to communicate using a serial channel with the master devices configured in the previous steps.

5. To send data interactively from the master device to the Arduino board or to see this data received in real time, we must use an application-type "serial terminal" as discussed in Chapter 3 for Windows, Linux and Mac OS X (if you are using a phone as a master device with Android, a good application is BlueTerm (https://play.google.com/store/apps). As always, all we need to do to use this type of programs is to select the serial port for the Bluetooth module (from the detected at that time, shown in a list) and the communication speed (in bits / s) appropriate. Once the master device data arrive, another issue would be what do with them (keep them in a weighted file or a database, or represent them graphically in real time, etc.).

If the RX and TX module lines remain connected and you want to reload a new sketch in Arduino, communication problems appear and the sketch may not be loaded correctly. One solution is to disconnect these two lines, load the sketch and re-connect. Another solution would be used instead of the RX / TX hardware, a pair of lines SoftwareSerial lines; in this way, the serial communication with the Bluetooth module does not interfere with the serial communication through USB.
8.10 Example: The following is an example code where how easy it is (after the above steps) Bluetooth management traffic between the Arduino board and a master device is shown. Specifically, this sketch makes Arduino master device to wirelessly send one message countless times every two seconds.

```
#include <SoftwareSerial.h>
// We connect the module to the pins 2 and 3 (of SoftwareSerial type)
SoftwareSerial bt (2.3);
void setup () {
bt.begin (115200); // Default speed modules BlueSMiRF
}
void loop () {
bt.println ("Hello");
delay (2000);
}
```

8.11 Example: The following sketch supposed to have an LED connected to digital output pin No. 13 of our Arduino. When this received through a Bluetooth signal (from a master device) the character "1", the LED will light; upon receipt of the character "0", the LED will turn off.

```
#include <SoftwareSerial.h> char character; SoftwareSerial bt (2.3);
void setup () {
bt.begin (115200);
pinMode (13, OUTPUT);
}
void loop () {
// While not received anything, I do nothing while (bt.available () == 0) {;}
bt.read character = ();
if (character == '0') digitalWrite (13, LOW);
if (character == '1') digitalWrite (13, HIGH);
delay (50);
}
```

163

The BlueSMiRF modules are sold with a set of values "factory" (bit / s by default, default key, default module name, etc.) that can be modified by us if we so desire. For this, we enter the "command mode" and send commands to your setup.

This is done by connecting the module through our application "serial terminal" preferred at a rate to 9600 bits / s and writing the special command "$$$" (without the quotes). Attention, this must be done within the first minute after the electric ignition module: then no longer be possible to enter the "command mode". If you consider this too short, you can modify and save the new time by exactly a given command.

Once written the "$$$" command, the module should respond with the message "CMD". From here, before sending any commands, we must first change the configuration of serial terminal program commands sent to end up on a character terminal (module requires it) so we need to set an option called "Newline "" Carriage return "or similar where it appeared before the" No line ending "or similar, which is what is usually set by default.

Once you have done this, and can send commands we want, which can be of three types: commands to obtain information module commands to set configuration parameters and command module to connect / disconnect with other modules. A command of the first type is for example "D" (without quotation marks), which identifies the basic data of the module (bits / s, name, address, etc.); one that offers more detailed information is the "E" command. A command of the second type can be the command "SP, nuevaclave" which serves to switch the device key (by default, "1234", remember) for the specified. A command of the third type can be "I, 30" command, which performs a scan for the specified seconds (30 in this case) to detect the presence of nearby Bluetooth devices. There are many more command, but in this book we have not enough space to deepen them; Fortunately, all commands are well documented both in the datasheet of RN-41 and RN-42 chips (in fact, they are the same commands for both) and the wonderful "Advanced User

Guide", downloadable from the product website in Sparkfun. To exit "command mode" and return to standard data transmission, it is to type the "---" (without quotes). Remember to set the "No line ending" option again.

Another interesting Bluetooth breakout board is called "JY-MCU BOARD BT" available at DealExtreme (among others) with product code number 104299. Work default to 9600 bits / s and your password is "1234". It also has four connectors: "VCC" (in this case connected to 5 V), "GND", "RXD" and "TXD" and has integrated PCB antenna. The steps are similar to those already described: load the code into the Arduino board, matching the module with Bluetooth devices
teachers to use, connected to the plate and use a serial terminal program to interact with it.

A breakout board Bluetooth curious is the "BTBee" of IteadStudio, as their shape and arrangement of the VCC, GND, RX and TX pins is compatible with XBee modules (and work, like them, to 3.3 V). This feature makes them ideal for when you have a system already fitted with XBee modules and desire to make a change to Bluetooth technology as quickly and easily as possible. It incorporates HC-06 module, which integrates an antenna on your PCB. This plate is acquired running default to 38400 bits / s with a key "0000". On the website the product is available many highly recommended to consult information: from the reference configuration command to several code examples, through different diagrams.

Another highlight is the plate "LC-05 Bluetooth Serial Module Master / Slave in One" manufactured by AliExpress and distributed among others ElectronDragon. As the name suggests, this board, unlike the previous ones, can function both as a slave (default) as a teacher (if it is configured using the appropriate commands). It can operate at 3.3 V or 5 V, works at 9600 bit / s and the default password is "1234".

Shields adding Bluetooth connectivity

To add Bluetooth connectivity, we can also use our attachable Arduino shields instead of separate modules. In this case, the way of working of the shields is very similar: through two pins electrically feed (5 V or 3V3 and GND) and through two other pins serial communication is set (software type normally not to interfere with existing serial communication hardware pins 0 and 1). These two pins can usually be chosen within our sketch.

One example is the "Bluetooth Shield" from Seeedstudio. This shield can act as both master and slave device, so we can serve both to communicate with a computer or mobile phone (as we have been seeing so far) but also for communication between two Bluetooth Arduinos with individual shields (one master and one slave). Because of this versatility, its configuration and use is more complex than seen so far. Keep in mind, however, that the maximum detection range is 10 meters (class 2). In the official wiki product offering extensive documentation on their default values, all possible configurations and several cases of use. In addition, a specific library for handling and several sample code itself provides. Other very similar to the previous shields (since they can also operate in master or slave mode) are "Bluetooth Shield" from ElecFreaks and "BT shield v2.2" of Iteadstudio, both based on the HC-05 module, also class 2 and with integrated antenna.

We must distinguish the "BT shield v2.2" from "BT shield v2.1" of Iteadstudio. The latter is a shield that is based HC-06 Bluetooth module (also class 2) but can only function as slave mode. This makes the configuration and operation of this shield easier. Specifically, the first thing to do is to get started on an Arduino board attach on but that is not running any code. To ensure this point, the easiest way is to upload a sketch in which the functions "setup ()" and "loop ()" are completely empty. With this we obtain power the shield so that it can be detected by the master device and can be matched (the default password is "1234"). Once recognized the shield by the master device, we must disengage the shield and then upload to the Arduino the desired code (via USB as always). This code must use the "Serial" object to the Bluetooth serial-communication (running at 9600 bits / s by default). Finally, we have to re-attach the shield on the Arduino board and ensure that the small switch is placed in this position incorporates "To Board".

If we set up the basic data of this shield (name, password, etc.), we will enter the "command mode". Therefore we should keep coupled to the Arduino and connect this to a computer via USB, placing the small switch on the shield "To FT232" position. We'll have to reload a completely empty sketch in Arduino and then execute our application "serial terminal" preferred at a rate of 9600 bits / s to connect the USB (not the Bluetooth!) Port: thereafter we can run the configuration commands you want. A useful command is by example "AT + PINxxxx" (without quotation marks), it used to change the key (the new location specified in the "x"). All available commands can be found on the datasheet of the HC-06 module. Once we're done, we disengage the shield and then upload to the Arduino the desired code (via USB as always). Finally, we have to re-attach the shield on the Arduino board, making sure the small switch is placed in this position incorporates "To Board".

## ARDUINO DISTRIBUTORS

### And electronic equipment

There are many online stores where you can buy different models of plates and Arduino shields. The price variations are usually a few euros and the overall service is good at all, so the choice of one or the other is usually based mainly on personal experience of each. Then they follow (much less extensively or completely) one of the most popular stores to buy everything you need for our projects with Arduino.

The first place we go is the official store of Arduino: http://store.arduino.cc. In addition to the plates and official shields, from here we can buy other elements, such as different unofficial shields, TinkerKit (separately or as a kit) modules, loose ATMEGA328P microcontroller but with the bootloader Optiboot already loaded, the controller chip L298N, microSD cards, PoE modules, USB cables of different types, strings of pin-pin-male and female in different lengths to place shields, engines etc. We can also buy basic electronic components to mount any circuit, such as resistors, potentiometers, LDRs, capacitors, LEDs, switches, cables of different lengths, clamps, breadboards of different sizes, etc.

Another site that is http://arduino.cc/en/Main/Buy can see where we can see an exhaustive list of the various official Arduino recognized dealers, classified by country. In all of them we can find different plates and official Arduino shields and other components needed to implement any project (from resistors, potentiometers, capacitors, diodes, LEDs, LCDs, transistors, buzzers, switches, breadboards, cables, multimeters ... to engines of various types, materials, robotics, TinkerKit modules, XBee modules, antennas, different types of power supplies and accessories, keypads, all kinds of chips, sensors of all kinds, plates and shields unofficial ...), also including the possibility of acquiring complete kits ready with the essentials. Some of the dealers listed in the "Spain" section are:

Electan (http://www.electan.com) CanaKit (http://www.canakit.es) Bricogeek (http://www.bricogeek.com/shop) Cooking-Hacks (http://www.cooking-hacks.com) Ardutienda (http://www.ardumania.es/ardutienda) Elect. Embaj. (Http://www.electronicaembajadores.com) Bcncybernetics (http://www.bcncybernetics.com)
Ro-Botica (http://www.ro-botica.com/arduino.asp), specializing in robotics.

There are also several distributors for Argentina, Brazil, Chile, Colombia, Costa Rica, Ecuador, Mexico, Panama and Uruguay.

But if we want to see for hours inmensísimo catalog presenting the major distributors of Arduino and electronics worldwide (and read many interesting articles and tutorials that very often offer about their products, some of which They are designed and manufactured by themselves), then a (far from exhaustive) list of the most important shows. Some of them have often appeared throughout the book.

Sparkfun (http://www.sparkfun.com) Adafruit (http://www.adafruit.com Makershed (http://www.makershed.com)

And also:

DFRobot (http://www.dfrobot.com) Freetronics (http://www.freetronics.com) Iteadstudio (http://iteadstudio.com/store) Seeedstudio (http://www.seeedstudio.com) Yourduino ( http://www.yourduino.com) Fungizmos (http://store.fungizmos.com) Modern Device (http://shop.moderndevice.com) Cutedigi (http://www.cutedigi.com) RSH Electronics (http://www.rshelectronics.co.uk) SK Pang (http://www.skpang.com) Littlebird Elect. (Http://littlebirdelectronics.com) Hacktronics (http://www.hacktronics.com) Hobbytronics (http://www.hobbytronics.co.uk) Mindkits (http://www.mindkits.com) Cool Components ( http://www.coolcomponents.co.uk) Oomlout (http://www.oomlout.co.uk)

Nkcelectronics (http://www.nkcelectronics.com)

EIO (http://www.eio.com)


Australian Robotics (http://australianrobotics.com.au)

Snootlab (http://www.snootlab.com)


Earthshine Design (http://www.earthshinedesign.com)

Kineteka (http://shop.kineteka.com)


EvilMadScience (http://evilmadscience.com)

Lees (http://www.leeselectronic.com)


Arduino providers that are more specifically specialized in robotics components are:


Pololu (http://www.pololu.com) Solarbotics (http://www.solarbotics.com) RobotShop (http://www.robotshop.com) RobotBits (http://robotbits.co.uk) TrossenRobotics ( http://www.trossenrobotics.com) SGBotic (http://www.sgbotic.com) ActiveRobots (http://www.active-robots.com) RobotGear (http://www.robotgear.com.au) Toysdownunder (http://www.toysdownunder.com) RobotStore (http://www.robotstore.com) RobotElectronics (http://www.robot-electronics.co.uk) SuperRobot (http: //www.superrobotica. com) Juguetronica (http://www.juguetronica.com)


Other suppliers worthy of mention are also: Open Electronics (http://store.open-electronics.org), specialized in remote control and location, Makerbot (http://store.makerbot.com) and RepRap (http: // reprap.org), specialized in the design and construction of 3D, or Microcontrollers Shop (http://microcontrollershop.com) printers, which sell everything related to Atmel and other manufacturers.

If we were looking for a strange electronic component that did not meet me at none of the above sites, after asking our nearest local store still could see the boundless catalogs of the following major suppliers worldwide of electrical and electronic components (the list is not exhaustive, far from it), including:

Mouser (http://es.mouser.com)
Jameco (http://www.jameco.com)

Farnell (http://es.farnell.com)
RS (http://es.rs-online.com)

Digikey (http://www.digikey.es)
Newark (http://www.newark.com)

Conrad (http://www.conrad.com)
Mpja (http://www.mpja.com)

Maplin (http://www.maplin.com) RadioShack (http://www.radioshack.com) Verical (http://www.verical.com) Vishay (http://www.vishay.com) Velleman (http://www.velleman.eu) Jaycar (http://www.jaycar.com) Satistronics ( http://www.satistronics.com) The Allied. (http://www.alliedelec.com) Future Electronics (http://www.futureelectronics.com) Futurlec (http://www.futurlec.com) Gateway Catalog (http://www.gatewaycatalog.com) BGMicro (http://www.bgmicro.com)
OnlineComponents (http://www.onlinecomponents.com)

As if that were not enough, we also have two very practical pages that can help us find out where a particular component sold and compare their prices (and actual stock) between different suppliers: http://www.findchips.com and http: // www.octopart.com.

Kits

However, the easiest for beginners is to acquire Arduino
called "kits", consisting of an Arduino (usually the model UNO) over a relatively
complete collection of electronic components and packaged. Some of them even
come with a textbook project step by step. Acquiring one of these kits, then the user
will not have to worry about looking each element individually need for their projects
because most of them have acquired and hit with the kit. Most projects seen in this
book can come true with any of these kits listed below:

"Arduino Starter Kit" (the official store of Arduino) contains resistors of different
values, a thermistor, an LDR, a potentiometer, a diode, capacitors of different kinds,
different color LEDs, switches, cables, one breadboard, pins for connecting shields,
a bipolar NPN transistor and one MOS, a buzzer a L293D chip, a 6.9 V DC motor
and a servo motor, a pair of tilt sensors, a pair of optocouplers and a USB cable A /
B to connect the computer to the plate. It also incorporates a book of 170 pages in
which 15 educational projects, from simple to more complicated are developed; the
complete list can be read in http://arduino.cc/en/Main/ArduinoStarterKit. There are
two variants of this kit: one that incorporates all of the above plus an Arduino UNO
board and another that does not.

"Getting started with Arduino kit" (from Makershed) contains resistors of different
values, a pair of LDRs a breadboard, cables, switches, LEDs of different colors, a
container 9 V battery connector of 2.1 mm, a USB cable and the Arduino UNO.
Another kit is the most comprehensive Makershed "Ultimate Microcontroller Pack"
which contains in addition to the above, two mini-servo motors, a DC minimotor, a
vibration motor, 16x2 LCD display, a sensitive resistance force ( FSR), a box for
storing the components, most models breadboards,
capacitors, tilt sensor, switches, thermistors, switches, a diode, a bipolar NPN
transistor, a speaker, a buzzer and several strings of pins.

"Inventor's kit" (Sparkfun) containing resistors of different values, potentiometers of various types, an LDR, diodes, LEDs of different colors, buttons, a small DC motor, a mini-servo motor, two bipolar NPN transistors, a relay , a temperature sensor, a bending sensor, a buzzer, cables, string pin a breadboard and a bracket to attach a USB cable and the Arduino UNO. It also contains an illustrated guide projects. A kit similar to the above is the "Starter Kit - Flex", which also contains a thermistor but does not incorporate or illustrated guide or transistors or diodes or engines or the relay or the temperature sensor, among other minor differences . On the other hand, SparkFun also distributes the "Jumper Wire Kit", which is simply an extra set of cables of different lengths, ideal to have them always on hand if missing in our prototyping projects.

"Starter Pack" (from Adafruit) contains different values resistors, potentiometers, an LDR, different colored LEDs, switches, cables, a small breadboard a Protoshield an AC / DC 9 V, a container for batteries 9 V with 2.1 mm connector, a USB cable and the Arduino UNO. Similar to the previous kit is the "Budget pack", but it contains neither the container for batteries or AC / DC adapter or the Protoshield, among other minor differences.

"ARDX Experimentation kit" (from Adafruit) contains different values resistors, potentiometers, one force sensitive resistor (FSR), an LSR, two diodes, LEDs of different colors, buttons, a small DC motor, a mini- servo, two bipolar NPN transistors, a relay, a temperature sensor, a buzzer, cables, a breadboard and a support for fixing a clip for attaching a 9 V battery to a 2.1 mm jack plug, a USB cable and Arduino UNO. It also contains an illustrated guide projects.

There are many other kits; noteworthy are the "Kit Workshop" Ardumania (whose most prominent components are a temperature sensor, an LDR, a mini-servo motor, a buzzer, a MOSFET transistor and a diode, but beware, does not include the cable USB or plate); the "Arduino Starter Kit", the "Arduino Sidekick Basic Kit" and "Arduino Lab Kit" Cooking-Hacks (each more complete, besides the

"Components Kit" that comes without Arduino); the "Starter Kit" Cutedigi (which can be purchased whole or by so-called "A", "B", "C", etc parts); the "Arduino Educational Kit" Hacktronics; "Low-cost Starter Set" Yourduino; the "Starter Kit" Earthshine (containing a manual very well written projects, like the "Arduino Workshop" SmileyMicros.com); the "Starter Kits" of TodoElectronica.com, the "Tools and Parts Electronics Starter Kit" CuriousInventor (without Arduino), the "Electronics Components Packs 1A" and "2a" of Makershed (without Arduino), the "Starter Kit Component "and" Resistor Starter Kit "Akafugu (without Arduino), etc.

If we are interested in the field of educational kits general electronics (no relation to Arduino), are interesting http://www.elektor.es web pages and said http://www.todoelectronica.com, which also They are responsible for teaching journals about exciting electronics.

Printable code

## TABLE OF ASCII

| Código numérico | Carácter | Código numérico | Carácter | Código numérico | Carácter |
|---|---|---|---|---|---|
| 33 | ! | 65 | A | 97 | a |
| 34 | " | 66 | B | 98 | b |
| 35 | # | 67 | C | 99 | c |
| 36 | $ | 68 | D | 100 | d |
| 37 | % | 69 | E | 101 | e |
| 38 | & | 70 | F | 102 | f |
| 39 | ´ | 71 | G | 103 | g |
| 40 | ( | 72 | H | 104 | h |
| 41 | ) | 73 | I | 105 | i |
| 42 | * | 74 | J | 106 | j |
| 43 | + | 75 | K | 107 | k |
| 44 | , | 76 | L | 108 | l |

| | | | | | |
|---|---|---|---|---|---|
| 4 | . | 7 | N | 11 | n |
| 4 | / | 7 | O | 11 | o |
| 4 | 0 | 8 | P | 11 | p |
| 4 | 1 | 8 | Q | 11 | q |
| 5 | 2 | 8 | R | 11 | r |
| 5 | 3 | 8 | S | 11 | s |
| 5 | 4 | 8 | T | 11 | t |
| 5 | 5 | 8 | U | 11 | u |
| 5 | 6 | 8 | V | 11 | v |
| 5 | 7 | 8 | W | 11 | w |
| 5 | 8 | 8 | X | 12 | x |
| 5 | 9 | 8 | Y | 12 | y |
| 5 | : | 9 | Z | 12 | z |
| 5 | ; | 9 | [ | 12 | { |
| 6 | < | 9 | \ | 12 | \| |
| 6 | = | 9 | ] | 12 | } |
| 6 | > | 9 | ^ | 12 | ~ |
| 6 | ? | 9 | | 12 | DEL |
| 6 | @ | 9 | ` | | |

If you want to know the full ASCII table (and its various extensions), you can consult http://www.asciitable.com. On Linux systems we can also view it using the command man ascii.